"A deeply researched, compulsively readable, subtly philosophical tour through the human brain . . . In just a few pages, Barrett dispels myths so deeply entrenched that many of us assumed they were indisputable scientific fact (goodbye, lizard brain!). She does this with the effortless concision of a poet, not a word wasted . . . [*Seven and a Half Lessons About the Brain*] deserves to be read and reread and then, just as important, to be thought about deeply."

— Daniel H. Pink, *New York Times* best-selling author of *Drive* and *When*

"Beautiful writing and sublime insights that will blow your mind like a string of firecrackers. If you want a rundown of the brain and its magic, start here."

— David Eagleman, Stanford neuroscientist, *New York Times* best-selling author of *Incognito* and *Livewired*

"*Seven and a Half Lessons About the Brain* reads like a novel—one whose main character is all of us. In fresh and lively prose, Barrett provides deep insight into what brains are for, how they operate and are programmed, how they create the 'reality' we experience, and how they ultimately produce our thoughts, feelings, and actions. Read this book! It will make you smarter about yourself and your species."

— Leonard Mlodinow, *New York Times* best-selling author of *The Drunkard's Walk, Subliminal,* and *Elastic*

"A radical and provocative look at a range of pervasive misconceptions, emerging discoveries, and enticing mysteries regarding our very nature as individuals and intertwined social beings. By illuminating our unimaginably complex, constantly changing brain/body networks, Barrett gets to the heart of the new understanding of who and what we are as creatures, and of how much latitude and agency we have." — Jon Kabat-Zinn, founder of Mindfulness-Based Stress Reduction (MBSR), author of *Full Catastrophe Living* and *The Healing Power of Mindfulness*

ALSO BY LISA FELDMAN BARRETT

How Emotions Are Made:
The Secret Life of the Brain

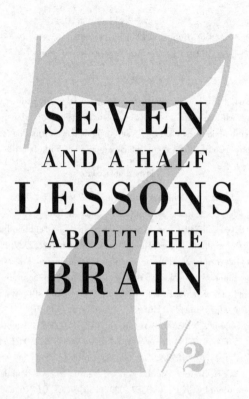

SEVEN
AND A HALF
LESSONS
ABOUT THE
BRAIN

Lisa Feldman Barrett

MARINER BOOKS
An Imprint of HarperCollins*Publishers*
Boston   New York

First Mariner Books edition 2021
Copyright © 2020 by Lisa Feldman Barrett
Illustrations by Flow Creative (flowcs.com)

marinerbooks.com

Library of Congress Cataloging-in-Publication Data
Names: Barrett, Lisa Feldman, author.
Title: Seven and a half lessons about the brain / Lisa Feldman Barrett.
Description: Boston : Houghton Mifflin Harcourt, 2020. |
Includes bibliographical references and index. | Summary: "From the author
of How Emotions Are Made, a myth-busting primer on the brain, in the
tradition of Seven Brief Lessons on Physics and Astrophysics
for People in a Hurry"—Provided by publisher.
Identifiers: LCCN 2020023431 (print) | LCCN 2020023432 (ebook) |
ISBN 9780358157144 (hardcover) | ISBN 9780358157120 (ebook) |
ISBN 9780358645597 (pbk.)
Subjects: LCSH: Brain—Popular works. | Neurosciences—Popular works.
Classification: LCC QP376 .B357 2020 (print) | LCC QP376 (ebook) |
DDC 612.8/2—dc23
LC record available at https://lccn.loc.gov/2020023431
LC ebook record available at https://lccn.loc.gov/2020023432

Book design by Emily Snyder

Droodles excerpted from *The Ultimate Droodles Compendium: The Absurdly Complete Collection of All the Classic Zany Creations of Roger Price.* Copyright © 2019 by Tallfellow Press, Inc. Used by permission. All rights reserved.

Printed in the United States of America
23 24 25 26 27 LBC 6 5 4 3 2

*To Barb Finlay
and my other colleagues who
taught me the craft of neuroscience,
for their great generosity and even
greater patience*

Contents

Author's Note

I wrote this book of short, informal essays to intrigue and entertain you. It's not a full tutorial on brains. Each essay presents a few compelling scientific nuggets about your brain and considers what they might reveal about human nature. The essays are best read in order, but you can also read them out of sequence.

As a professor, I usually include loads of scientific details in my writing, such as descriptions of studies and pointers to journal papers. For these informal essays, however, I've moved the full scientific references to my website, sevenandahalflessons.com.

Also, at the end of this book, you'll find an appendix with selected scientific details. It offers a bit more depth on some essay topics, explains that certain points are still debated by scientists, and gives credit to other people for some interesting turns of phrase.

Why are there seven and a half lessons rather than

eight? The opening essay tells a story of how brains evolved, but it is just a brief peek into a vast evolutionary history — hence, half a lesson. The concepts that it introduces are critical to the rest of the book.

I hope you'll enjoy learning what one neuroscientist thinks is fascinating about your brain and how that three-pound blob between your ears makes you human. The essays don't tell you what to think about human nature, but they do invite you to think about the kind of human you are or want to be.

The Half-Lesson

Your Brain Is Not for Thinking

Once upon a time, the Earth was ruled by creatures without brains. This is not a political statement, just a biological one.

One of these creatures was the amphioxus. If you ever glimpsed one, you'd probably mistake it for a little worm until you noticed the gill-like slits on either side of its body. Amphioxi populated the oceans about 550 million years ago, and they lived simple lives. An amphioxus could propel itself through the water, thanks to a very basic system for movement. It also had an exceedingly simple way of eating: it planted itself in the seafloor, like a blade of grass, and consumed any minuscule creatures that happened to drift into its mouth. Taste and smell were of no concern because an amphioxus didn't have senses like yours. It had no eyes, just a few cells to detect changes in light, and it could not hear. Its meager nervous system included a teeny clump of cells that was not quite a brain. An amphioxus, you could say, was a stomach on a stick.

Amphioxi are your distant cousins, and they're still around today. When you look at a modern amphioxus, you behold a creature very similar to your own ancient, tiny ancestor who roamed the same seas.

Can you picture a little wormy creature, two inches long, swaying in the current of a prehistoric ocean, and glimpse humanity's evolutionary journey? It's difficult. You have so much that the ancient amphioxus did not: a few hundred bones, an abundance of internal organs, some limbs, a nose, a charming smile, and, most important, a brain. The amphioxus didn't need a brain. Its cells for sensing were connected to its cells for moving, so it reacted to its watery world without much processing. You, however, have an intricate, powerful brain that gives rise to mental events as diverse as thoughts, emotions, memories, and dreams—an internal life that shapes so much of what is distinctive and meaningful about your existence.

Why did a brain like yours evolve? The obvious answer is *to think*. It's common to assume that brains evolved in some kind of upward progression—say, from lower animals to higher animals, with the most sophisticated, thinking brain of all, the human brain, at the top. After all, thinking is the human superpower, right?

Well, the obvious answer turns out to be wrong. In fact, the idea that our brains evolved for thinking has been the source of many profound misconceptions about human

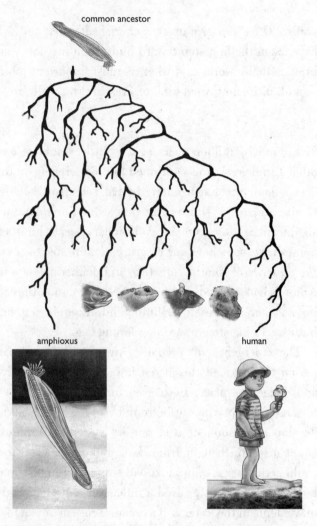

Amphioxi were not our direct ancestors, but we had a common ancestor that was very likely similar to a modern-day amphioxus.

nature. Once you give up that cherished belief, you will have taken the first step toward understanding how your brain actually works and what its most important job is —and, ultimately, what kind of creature you really are.

⟨⟩

Five hundred million years ago, as little amphioxi and other simple creatures continued to dine serenely on the ocean floor, the Earth entered what scientists call the Cambrian period. During this time, something new and significant appeared on the evolutionary scene: hunting. Somewhere, somehow, one creature became able to *sense the presence* of another creature and deliberately ate it. Animals had gobbled one another before, but now the eating was more purposeful. Hunting didn't require a brain, but it was a big step toward developing one.

The emergence of predators during the Cambrian period transformed the planet into a more competitive and dangerous place. Both predators and prey evolved to sense more of the world around them. They began to develop more sophisticated sensory systems. Amphioxi could distinguish light from dark, but newer creatures could actually see. Amphioxi had simple skin sensation, but newer creatures evolved a fuller sense of their body movements in the water and a greater sense of touch that allowed them to detect objects by vibration. Sharks today still use this kind of touch sense to locate prey.

With the arrival of greater senses, the most critical

question in existence became *Is that blob in the distance good to eat, or will it eat me?* Creatures who could better sense their surroundings were more likely to survive and thrive. The amphioxus may have been a master of its environment, but it couldn't sense that it *had* an environment. These new animals could.

The hunters and the hunted also received a boost from another new ability: more sophisticated kinds of movement. For the amphioxus, whose nerves for sensing and moving were woven together, movement was extremely basic. Whenever its stream of food became a trickle, it wriggled in a random direction to plant itself in another spot. Any looming shadow prompted its body to dart away. In the new world of hunting, however, predators and prey alike began to evolve more capable systems for movement, or motor systems, to navigate with greater speed and dexterity. These newer animals could dart, turn, and dive deliberately toward things like food and away from things like threats in ways that suited their environment.

Once creatures could sense at a distance and make more sophisticated movements, evolution favored those who performed these tasks efficiently. If they chased a meal but moved too slowly, something else caught the meal and ate it first. If they burned up energy fleeing from a potential threat that never arrived, they wasted resources that they might have needed later. Energy efficiency was a key to survival.

You can think about energy efficiency like a budget. A

financial budget tracks money as it's earned and spent. A budget for your body similarly tracks resources like water, salt, and glucose as you gain and lose them. Each action that spends resources, such as swimming or running, is like a withdrawal from your account. Actions that replenish your resources, such as eating and sleeping, are like deposits. This is a simplified explanation, but it captures the key idea that running a body requires biological resources. Every action you take (or don't take) is an economic choice—your brain is guessing when to spend resources and when to save them.

The best way to keep to a financial budget, as you may know from personal experience, is to avoid surprises—to anticipate your financial needs before they arise and make sure you have the resources to meet them. The same is true of a body budget. Little Cambrian creatures needed an energy-efficient way to survive when a hungry predator was nearby. Should they wait around until the ravenous beast made its move and then react by freezing or hiding? Or should they anticipate the lunge and prepare their bodies in advance to escape?

When it came to body budgeting, prediction beat reaction. A creature that prepared its movement before the predator struck was more likely to be around tomorrow than a creature that awaited a predator's pounce. Creatures that predicted correctly most of the time, or made nonfatal mistakes and learned from them, did well. Those that frequently predicted poorly, missed threats, or false-

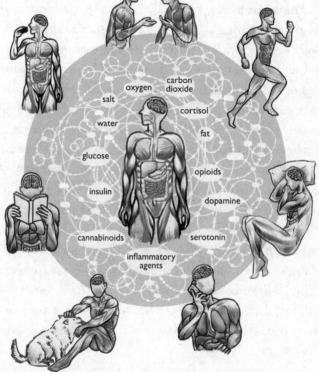

Your brain runs a budget for your body that regulates water, salt, glucose, and many other biological resources inside you. Scientists call the budgeting process *allostasis*.

alarmed about threats that never materialized didn't do so well. They explored their environment less, foraged less, and were less likely to reproduce.

The scientific name for body budgeting is *allostasis*. It means automatically predicting and preparing to meet the body's needs *before they arise*. As Cambrian creatures acquired and spent resources throughout the day by sensing and moving, allostasis kept the systems of their bodies in balance most of the time. Withdrawals were fine, as long as they renewed the spent resources in a timely manner.

How can animals predict their bodies' future needs? The best source of information comes from their past — the actions they've taken at other times in similar circumstances. If a past action brought benefits, such as a successful escape or a tasty meal, they're likely to repeat that action. All sorts of animals, including humans, somehow conjure up past experiences to prepare their bodies for action. Prediction is such a useful capability that even single-celled creatures plan their actions predictively. Scientists are still puzzling out how they do it.

So imagine a tiny Cambrian creature drifting in the current. Up ahead, it senses an object that might be tasty to eat. What now? It can move, but should it? After all, moving takes energy from the budget. The movement should be *worth the effort*, economically speaking. *That* is a prediction, based on past experience, to prepare a body for action. To be clear, I don't mean a conscious, thoughtful decision, weighing the pros and cons. I'm saying that *something* must occur inside a creature to predict and launch one set of movements rather than another. That *something* reflects a determination of worth. The value of any movement is intimately bound up with body budgeting by allostasis.

Meanwhile, ancient animals continued to evolve larger, more complex bodies. That meant the insides of bodies became more sophisticated. The amphioxus, the little stomach on a stick, had almost no bodily systems to regulate. A handful of cells were enough to keep its body upright in the water and digest food within its primitive gut. Newer animals, however, developed intricate internal systems, like a cardiovascular system with a heart that pumps blood, a respiratory system that takes in oxygen and eliminates carbon dioxide, and an adaptable immune system that fights infection. Systems like these made body budgeting much more challenging, less like a single bank account and more like the accounting department of a sizable company. These complex bodies needed something more than a handful of cells to ensure that water and blood and salt and oxygen and glucose and cortisol and sex hormones and dozens of other resources were all regulated well to keep a body running efficiently. They needed a command center. A *brain*.

And so, as animals gradually evolved bigger bodies with more systems to maintain, their handful of body-budgeting cells also evolved to become brains of greater and greater complexity. Fast-forward a few hundred million years, and the Earth is now littered with complicated brains of all kinds, including yours—a brain that efficiently supervises over six hundred muscles in motion, balances dozens of different hormones, pumps blood at a rate of two thousand gallons per day, regulates the energy of billions of brain cells, digests food, excretes waste, and fights illness, all of it nonstop for seventy-two years, give or take. Your body budget is like thousands of financial accounts in a giant, multinational corporation, and you have a brain that's up to the task. And all your body budgeting takes place in a massively complicated world made even more challenging by the other brains-in-bodies that you share it with.

So, returning to our original question: Why did a brain like yours evolve? That question is not answerable because evolution does not act with purpose—there is no "why." But we *can* say what is your brain's most important job. It's not rationality. Not emotion. Not imagination, or creativity, or empathy. Your brain's most important job is to control your body—to manage allostasis—by predicting energy needs before they arise so you can efficiently make worthwhile movements and survive. Your brain continually invests your energy in the hopes of earning a good return, such as food, shelter, affection, or physical protection, so you can perform nature's most vital task: passing your genes to the next generation.

In short, your brain's most important job is not thinking.

It's running a little worm body that has become very, very complicated.

Of course, your brain *does* think and feel and imagine and create hundreds of other experiences, such as letting you read and understand this book. But all of these mental capacities are consequences of a central mission to keep you alive and well by managing your body budget. Everything your brain creates, from memories to hallucinations, from ecstasy to shame, is part of this mission. Sometimes your brain budgets for the short term, like when you drink coffee to stay up late and finish a project, knowing that you are borrowing energy that you'll pay for tomorrow. Other times, your brain budgets for the long term, like when you spend years to learn a difficult skill, such as math or carpentry, that requires a sustained investment but ultimately helps you survive and prosper.

You and I do not experience our every thought, every feeling of happiness or anger or awe, every hug we give or receive, every kindness we extend, and every insult we bear as a deposit or withdrawal in our metabolic budgets, but under the hood, that is what's happening. This idea is key to understanding how your brain works and, in turn, how to stay healthy and live a longer and more meaningful life.

This little evolutionary story is the beginning of a longer tale about your brain and the other brains around you. In the next seven short lessons, we'll take a tour of remarkable scientific findings in neuroscience, psychology, and anthropology that have revolutionized our understanding of what happens inside your skull. You'll learn what makes the human brain distinctive in an animal kingdom full of astonishing brains. You'll explore how infant brains gradually transform into adult brains. And you'll discover how different kinds of human minds can arise from a single human brain structure. We'll even tackle the question of reality: What gives us the power to invent customs, rules, and civilizations? Along the way, we'll revisit body budgeting and prediction and their central roles in creating your actions and your experiences. We'll also uncover the powerful connections between your brain, your body, and other human brains-in-bodies. By the end of this book, I hope you will delight, as I do, in knowing that your thinking cap is for much more than thinking.

Lesson No.

1

You Have One Brain (Not Three)

Two thousand years ago, in ancient Greece, a philosopher named Plato recounted a war. Not a war between cities or nations but inside of each human being. Your human mind, wrote Plato, is a never-ending battle between three inner forces to control your behavior. One force consists of basic survival instincts, like hunger and sex drive. The second force consists of your emotions, such as joy, anger, and fear. Together, Plato wrote, your instincts and emotions are like animals that can pull your behavior in divergent, perhaps ill-advised directions. To counteract this chaos, you have the third inner force, rational thought, to rein in both beasts and guide you on a more civilized and righteous path.

Plato's compelling morality tale of inner conflict remains one of the most cherished narratives in Western civilization. Who among us has never felt an inner tug-of-war between desire and reason?

Perhaps it's unsurprising, then, that scientists later mapped Plato's battle onto the brain in an attempt to explain how the human brain evolved. Once upon a time, they said, we were lizards. Three hundred million years ago, that reptilian brain was wired for basic urges like feeding, fighting, and mating. About one hundred million years later, the brain evolved a new part that gave us emotions; then we were mammals. Finally, the brain evolved a rational part to regulate our inner beasts. We became human and lived logically ever after.

According to this evolutionary story, the human brain ended up with three layers—one for surviving, one for feeling, and one for thinking—an arrangement known as the *triune brain*. The deepest layer, or lizard brain, which we allegedly inherited from ancient reptiles, is said to house our survival instincts. The middle layer, dubbed the *limbic system*, supposedly contains ancient parts for emotion that we inherited from prehistoric mammals. The outermost layer, part of the cerebral cortex, is said to be uniquely human and the source of rational thought; it's known as the *neocortex* ("new cortex"). One part of your neocortex, called the prefrontal cortex, supposedly regulates your emotional brain and your lizard brain to keep your irrational, animalistic self in check. Advocates of the triune brain note that humans have a very large cerebral cortex, which they see as evidence of our distinctly rational nature.

You might have noticed that I've now offered two dif-

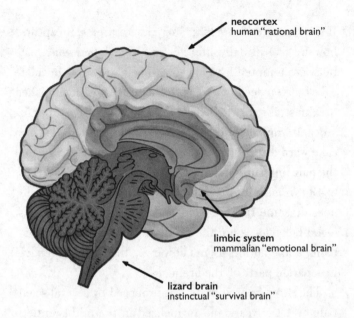

The triune brain idea

ferent descriptions of the evolution of the human brain. In the earlier half-lesson, I wrote that brains evolved increasingly elaborate sensory and motor systems while budgeting the energy resources of increasingly complex bodies. But the triune brain story says the brain evolved in layers that allow rationality to conquer our animalistic urges and emotions. How can we reconcile these two scientific views?

Fortunately, we don't have to reconcile them, because one of them is wrong. The triune brain idea is one of the most successful and widespread errors in all of science.

It's certainly a compelling story, and at times, it captures how we feel in daily life. For example, when your taste buds are tempted by a luscious slice of velvety chocolate cake but you decline it because, honestly, you just finished breakfast, it's easy to believe that your impulsive inner lizard and your emotional limbic system pushed you in a cake-ward direction, and your rational neocortex wrestled the pair into submission.

But *human brains don't work that way.* Bad behavior doesn't come from ancient and unbridled inner beasts. Good behavior is not the result of rationality. And rationality and emotion are not at war . . . they do not even live in separate parts of the brain.

The three-layered brain was proposed by several scientists over the years and formalized in the mid-twentieth century by a physician named Paul MacLean. He envisioned a brain that was structured like Plato's battle and confirmed his hypothesis using the best technology available at the time: visual inspection. That meant peering through a microscope at the brains of various dead lizards and mammals, including humans, and identifying their similarities and differences by sight alone. MacLean determined that the human brain had a collection of new parts that other mammal brains didn't, which he called the neocortex. He also concluded that mammal brains had a collection of parts that reptile brains didn't, which he called the limbic system. And voilà, a human origin story was born.

MacLean's tale of the triune brain gained traction in certain sectors of the scientific community. His speculations were simple, elegant, and seemingly consistent with Charles Darwin's ideas about the evolution of human cognition. Darwin asserted, in his book *The Descent of Man*, that the human mind had evolved along with the body, and therefore each of us harbors an ancient inner beast that we tame through rational thought.

The astronomer Carl Sagan introduced the idea of the triune brain to the wider public in 1977 in his book *The Dragons of Eden*, which won a Pulitzer Prize. Today, terms like *lizard brain* and *limbic system* run rampant through popular-science books and newspaper and magazine articles. While writing this lesson, in fact, I came across a special issue of *Harvard Business Review* in my local supermarket that explained how to "stimulate your customer's lizard brain to make a sale." Beside it sat a special issue of *National Geographic* that listed brain regions that make up the alleged "emotional brain."

What's less known is that *The Dragons of Eden* appeared when experts in brain evolution already had strong evidence that the triune brain story was incorrect: evidence hidden from the naked eye, within the molecular makeup of brain cells called neurons. By the 1990s, experts had completely rejected the idea of a three-layered brain. It simply didn't hold up when they analyzed neurons with more sophisticated tools.

In MacLean's day, scientists compared one animal

brain to another by injecting them with dye, slicing them paper-thin like deli meat, and squinting at the stained slices through a microscope. Neuroscientists who study brain evolution today still do this, but they also use newer methods that allow them to peer inside neurons and examine the genes within. They've discovered that neurons from two species of animals can *look* very different but *still contain the same genes*, suggesting that those neurons have the same evolutionary origin. If we find the same genes in certain human and rat neurons, for example, then similar neurons with those genes were most likely present in our last common ancestor.

Using these methods, scientists have learned that evolution does not add layers to brain anatomy like geological layers of sedimentary rock. But human brains are obviously different from rat brains, so how exactly did our brains come to differ if not by adding layers?

It turns out that as brains become larger over evolutionary time, they reorganize.

Let me explain with an example. Your brain has four clusters of neurons, or brain regions, that allow you to sense your body movements and help create your sense of touch. These brain regions are collectively called the primary somatosensory cortex. In a rat brain, however, the primary somatosensory cortex is just a single region that performs the same tasks. If we simply inspected human and rat brains by eye, as MacLean did, we might come to believe that rats lack three somatosensory regions found in the human brain. We might therefore conclude that these three regions are newly evolved in humans and must have new, human-specific functions.

Scientists have found, however, that your four regions and the rat's single region contain many of the same genes. This scientific tidbit suggests something about evolution; namely, the last common ancestor of humans and rodents, which lived about sixty-six million years ago, probably had a single somatosensory region that carried out some functions that our four regions do today. The single region most likely expanded and subdivided to redistribute its responsibilities as our ancestors evolved larger brains and bodies. This arrangement among brain regions — segregating and then integrating — creates a more complex brain that can control a larger and more complex body.

It's a tricky business to compare brains of different species to discover what is similar, because the path of evolution is twisty and unpredictable. What you see is not always what you get. Parts that look different to the naked eye can be similar genetically, and parts that differ genetically can look very similar. And even if you do find the same genes in the brains of two different animals, those genes can have different functions.

Thanks to recent research in molecular genetics, we now know that reptiles and nonhuman mammals have the same kinds of neurons that humans do, even those neurons that create the fabled human neocortex. Human brains did not emerge from reptile brains by evolving

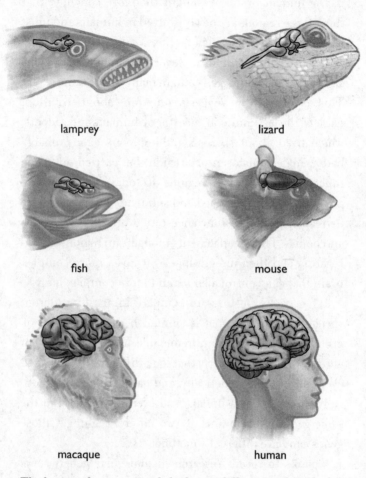

lamprey

lizard

fish

mouse

macaque

human

The brains of many animals look very different to the naked eye.

extra parts for emotion and rationality. Instead, something more interesting happened.

Scientists have recently discovered that the brains of *all mammals* are built from a single manufacturing plan, and most likely, the brains of reptiles and other vertebrates follow that same plan. Many people, including many neuroscientists, are not familiar with this work, and those who know about it are only beginning to reckon with its implications.

The common brain-manufacturing plan begins shortly after conception, when an embryo starts producing neurons. The neurons that form a mammal's brain are created in an astonishingly predictable order. The ordering holds true for mice, rats, dogs, cats, horses, anteaters, humans, and every other mammalian species studied so far, and genetic evidence strongly suggests the order holds for reptiles, birds, and some fish. Yes, to the best of our scientific knowledge, you have the same brain plan as a bloodsucking lamprey.

If the brains of so many vertebrates develop in the same order, why do these brains look so different from one another? Because the manufacturing process runs in stages, and the stages last for *shorter or longer durations* in different species. The biological building blocks are the same; what differs is the timing. For example, the stage that produces neurons for the cerebral cortex in humans runs for a shorter time in rodents and a much shorter time in lizards, so your cerebral cortex is large, a mouse's is smaller, and an iguana's is tiny (or nonexistent—it's debatable). If you could magically reach into a lizard embryo and force that stage to run for as long as it does in humans, it would produce something like a human cerebral cortex. (Though it wouldn't function like a human one. Size isn't everything, even for a brain.)

So the human brain has no new parts. The neurons in your brain can be found in the brains of other mammals and, likely, other vertebrates. This discovery undermines the evolutionary foundations of the triune brain story.

What about the rest of the story, that the human brain has an unusually large cerebral cortex that makes us the most rational animal? Well, it's true that our cerebral cortex is big and has expanded over evolutionary time, and that allows us to do certain things a bit better than other animals, as we'll learn in later lessons. But the real question here is whether the human cerebral cortex has gotten bigger, proportionally speaking, relative to the rest of the brain. So it's more scientifically meaningful to ask: Is our cerebral cortex unusually large *given our overall brain size?*

To understand why this is a better question, let's consider an analogy. Think for a moment about the variety of kitchens that you've seen in people's homes. Some kitchens are large and some are small. Imagine that you find yourself inside a gigantic kitchen. You might think, *Wow, these people must love to cook.* Is this a reasonable conclusion? No, not based on the kitchen size alone. You must also consider the kitchen in proportion to the rest of the house. A big kitchen in a big house is ordinary — it's just a scaled-up version of a typical house plan. A huge kitchen in a small house, however, is much more likely to have a special reason for its size, like the occupants are gourmet chefs.

The same principle applies to brains. A big brain with a proportionally big cerebral cortex would not be special, and, in fact, that's exactly what we humans have. *All* mammals have a relatively big cortex in a brain that's relatively large for their body size. Our cortex is just a scaled-up version of the relatively smaller cortex found in relatively smaller-brained monkeys, chimps, and many carnivores. It's also a scaled-down version of the larger cortex found in the larger brains of elephants and whales. If a monkey's brain could grow to human size, its cerebral cortex would be the same size as ours. Elephants have much more cerebral cortex than we do, but so would an elephant-sized human brain.

The size of our cerebral cortex, therefore, is not evolutionarily novel and does not require any special explanation. The size also says nothing about how rational a species is. (If it did, our most famous philosophers might be Horton, Babar, and Dumbo.) Western scientists and intellectuals concocted the idea of the big, rational cortex and have kept it alive for many years. The real story is that during the course of evolution, certain genes mutated to cause particular stages of brain development to run for longer or shorter times, producing a brain with proportionally bigger or smaller parts.

So you don't have an inner lizard or an emotional beast-brain. There is no such thing as a limbic system dedicated to emotions. And your misnamed neocortex is not a new part; many other vertebrates grow the same neurons that, in some animals, organize into a cerebral cortex if key stages run for long enough. Anything you read or hear that proclaims the human neocortex, cerebral cortex, or prefrontal cortex to be the root of rationality, or says that the frontal lobe regulates so-called emotional brain areas to keep irrational behavior in check, is simply outdated or woefully incomplete. The triune brain idea and its epic battle between emotion, instinct, and rationality is a modern myth.

To be clear, I'm not saying that our big brain has no advantages. (What advantages does it provide? The answers will unfold in the lessons that follow.) And while it's true that we're the only animal that can build skyscrapers and invent French fries, these abilities are not due to our big brains alone, as we shall see. Moreover, other animals have evolved abilities that surpass ours in significant ways. We don't have wings to fly. We can't lift fifty times our own weight. We can't regrow amputated body parts. Such abilities are superhero powers to us but business as usual for allegedly lesser creatures. Even bacteria are more talented than we are at certain tasks, like surviving in harsh, unfamiliar environments such as outer space or the insides of your intestines.

Natural selection did not aim itself toward us — we're just an interesting sort of animal with particular adaptations that helped us survive and reproduce in particular environments. Other animals are not inferior to humans. They are uniquely and effectively adapted to their environments. Your brain is not *more* evolved than a rat or lizard brain, just *differently* evolved.

If that's the case, why is the myth of the triune brain still popular? Why do college textbooks still depict a limbic system in the human brain and say it's regulated by the cerebral cortex? Why do expensive executive-training courses teach CEOs to get a grip on their lizard brains if experts in brain evolution dismissed such ideas decades ago? Partly it's because those experts need a better public relations department. But mostly it's because the triune brain is a story that comes with its own cheering section. With our unique capacity for rational thought, the story goes, we triumphed over our animal nature and now rule the planet. To believe in the triune brain is to award ourselves a first prize trophy for Best Species.

The idea of Plato's war, with rationality versus emotion and instinct, has long been Western culture's best explanation for our behavior. If you restrain your instincts and emotions appropriately, then your behavior is said to be rational and responsible. If you choose not to act rationally, then your behavior may be called immoral, and if you're unable to act rationally, you are considered mentally ill.

But what is rational behavior, anyway? Traditionally, it's the absence of emotion. Thinking is viewed as rational, whereas emotion is supposedly irrational. But that isn't necessarily so. Sometimes emotion is rational, like when you feel afraid because you're in imminent danger. And sometimes thinking isn't rational, like when you scroll through social media for hours, telling yourself you're bound to come across something important.

Perhaps rationality is better defined in terms of the brain's most important job: body budgeting—managing all the water, salt, glucose, and other bodily resources we use every day. In this view, rationality means spending or saving resources to succeed in your immediate environment. Let's say you're in a physically dangerous situation, and your brain prepares you to flee. It directs your adrenal glands, which sit atop your kidneys, to pump you full of cortisol, a hormone that provides a quick burst of energy. From a triune brain perspective, the cortisol rush is instinctual, not rational. But from a body-budgeting perspective, the cortisol rush is rational, because your brain is making a sound investment in your survival and the existence of your potential offspring.

If there was no danger and your body prepared to flee anyway, would that be irrational behavior? It depends on context. Suppose you're a soldier in a war zone, where threats appear on a regular basis. It's appropriate for your brain to frequently predict threat. It may sometimes guess incorrectly and flood you with cortisol when there's no danger. In one sense, we could view this false alarm as needless spending of resources that you may need later and therefore irrational. But in a war zone, this false alarm may be rational from a body-budgeting standpoint. You might waste a bit of glucose or other resources in the moment, but over the long run, you are more likely to survive.

If you return home from war to a safer environment but your brain continues to false-alarm, as happens in post-traumatic stress disorder, that behavior could still be considered rational. Your brain is protecting you from threats it believes are present, even though the frequent withdrawals decimate your body budget. The problem is your brain's beliefs; they are not a good fit for your new environment, and your brain hasn't adjusted yet. What we call mental illnesses, then, may be rational body-budgeting for the short term that's out of sync with the immediate environment, the needs of other people, or your own best interests down the road.

Rational behavior, therefore, means making a good body-budgeting investment in a given situation. When you exercise vigorously, you may have a rush of cortisol in your bloodstream and you may feel unpleasant, but we'd consider exercise rational because it's beneficial for your future health. The surge of cortisol when you receive criticism from a coworker might also be rational because it makes more glucose available so you can learn something new.

These ideas, if taken seriously, could shake the foundations of all sorts of sacred institutions in our society. In the law, for example, attorneys plead that their clients' emotions overwhelmed their reason in the heat of passion, and therefore they aren't fully to blame for their actions. But feeling distressed is not evidence of being irrational or that your so-called emotional brain has hijacked your supposed rational brain. Distress can be evidence that your whole brain is expending resources toward an anticipated payoff.

Many other social institutions are steeped in the idea of a mind at war with itself. In economics, models for investor behavior assume a sharp distinction between the rational and the emotional. In politics, we have leaders with clear conflicts of interest, such as past lobbying work in industries that they now oversee, who believe they can easily set aside their emotions and make rational decisions for the good of the people. Beneath these lofty ideas lurks the myth of the triune brain.

You have one brain, not three. To move past Plato's ancient battle, we might need to fundamentally rethink what it means to be rational, what it means to be responsible for our actions, and perhaps even what it means to be human.

Lesson No.

Your Brain Is a Network

THE BRAINS ON THIS PLANET have been pondering brains for thousands of years. Aristotle believed the brain was a cooling chamber for the heart, sort of like the radiator in your car. Philosophers in the Middle Ages maintained that certain brain cavities housed the human soul. In the nineteenth century, a popular idea called phrenology portrayed the brain as a jigsaw puzzle, where each piece produced a different human quality, such as self-esteem, destructiveness, or love.

A cooling chamber, a house for the soul, a jigsaw puzzle — these are all just metaphors invented to help us understand what brains are and how they work.

Today, we remain surrounded by so-called facts about the brain that are also just metaphors. If you've heard that the left side of your brain is logical and the right side is creative, that's just a metaphor. So is the idea that your brain has a "System 1" for quick, instinctive responses

and a "System 2" for slower, more thoughtful processing, concepts discussed in the book *Thinking, Fast and
Slow* by psychologist Daniel Kahneman. (Kahneman is
very clear that Systems 1 and 2 are metaphors about the
mind; but they are often mistaken for brain structures.)
Some scientists describe the human mind as a collection
of "mental organs" for fear, empathy, jealousy, and other
psychological tools that evolved for survival, but the brain
itself isn't structured like that. Your brain also does not
"light up" with activity, as if some parts are on and others off. It does not "store" memories like computer files
to be retrieved and opened later. These ideas are metaphors that emerged from beliefs about the brain that are
now outdated.

If real brains don't work like any of these metaphors
suggest they do, and the triune brain is a myth, then what
kind of brain do we actually have that makes us the kind
of animal that we are? What kind of brain gives us our
ability to cooperate, our capacity for language, and our
talent to guess what other people are thinking or feeling? What kind of a brain is *necessary* to create a human
mind?

The answer to these questions begins with an important insight. Your brain is a *network*—a collection of
parts that are connected to function as a single unit. You
are surely familiar with other networks that surround us.
The internet is a network of connected devices. An ant-
hill is a network of underground locations connected

by tunnels. Your social network is a collection of connected people. Your brain, in turn, is a network of 128 billion neurons connected as a single, massive, and flexible structure.

A brain network is not a metaphor. It's a description that comes from the best available science about how brains evolved, how they're structured, and how they function. And as you will see, this network structure will take us one step closer to understanding what makes your brain able to create your mind.

How do 128 billion individual neurons become a single brain network? Generally speaking, each neuron looks like a little tree, with bushy branches at the top, a long trunk, and roots at the bottom. (Yes, I know, I'm using a metaphor!) The bushy branches, which are called dendrites, receive signals from other neurons, and the trunk, which is called the axon, sends signals to other neurons through its roots.

Your 128 billion neurons continually fire off communications to each other, day and night. When a neuron fires, an electrical signal races down its trunk to its roots. This signal causes the roots to release chemicals into the gaps between neurons, called synapses. The chemicals travel across synapses and attach to another neuron's bushy top, causing that neuron to fire as well, and voilà, one neuron has passed information to another.

This arrangement of dendrites, axons, and synapses knits your 128 billion individual neurons into a network.

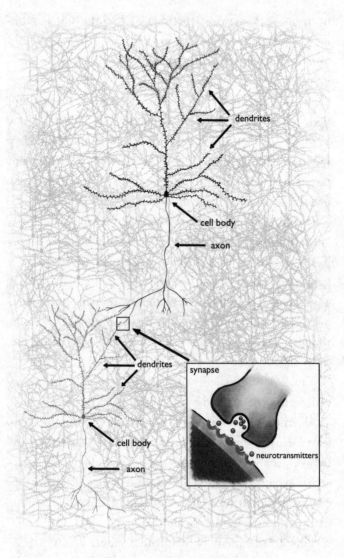

Neurons and their wiring

To make things simpler, I'll refer to this whole arrangement as the "wiring" of your brain.

Your brain network is always on. Your neurons never just sit around waiting for something in the outside world to make them fire. Instead, all of your neurons chat constantly with one another through their wiring. Their communications may become stronger or weaker depending on what's happening in the world and in your body, but the conversation never stops until you die.

Communication in your brain is a balancing act between speed and cost. Each neuron directly passes information to just a few thousand other neurons and receives information from a few thousand others, give or take, yielding over five hundred trillion neuron-to-neuron connections. That's a really big number, but it would be considerably larger if every neuron spoke directly to *every* other neuron in the network. Such a structure would require so many more connections that your brain would run out of resources to sustain itself.

So you have a more frugal wiring arrangement that is sort of like the global air-travel system. (Yep, here comes another metaphor.) The air-travel system is a network of about seventeen thousand airports around the world. Whereas your brain carries electrical and chemical signals, this network carries passengers (and, if we're lucky, our luggage). Each airport runs direct flights to *some* other airports but not to *every* other airport. If every airport sent flights to every other, air traffic would increase by billions more flights per year, and the whole system would run out of fuel and pilots and runways and ultimately collapse. Instead, some airports take the burden off the rest by serving as hubs. There's no direct flight from Lincoln, Nebraska, to Rome, Italy, so you first fly from Lincoln to a hub like Newark International Airport in New Jersey, then hop onto a second, longer-distance flight from the hub to Rome. You might even take three flights and pass through two hubs on your journey. The hub system is flexible and scalable, and it forms the backbone of international travel. It allows all airports to participate globally, even while many of them focus on local flights.

Your brain network is organized in much the same way. Its neurons are grouped into clusters that are like airports. Most of the connections in and out of a cluster are local, so, like an airport, the cluster serves mostly local traffic. In addition, some clusters serve as hubs for communication. They are densely connected to many other clusters, and some of their axons reach far across the brain and act as long-distance connections. Brain hubs, like airport hubs, make a complicated system efficient. They allow most neurons to participate globally even as they focus more locally. Hubs form the backbone of communication throughout the brain.

Hubs are supercritical infrastructure. When a major airport hub like Newark or London's Heathrow goes down, flight delays and cancellations ripple across the

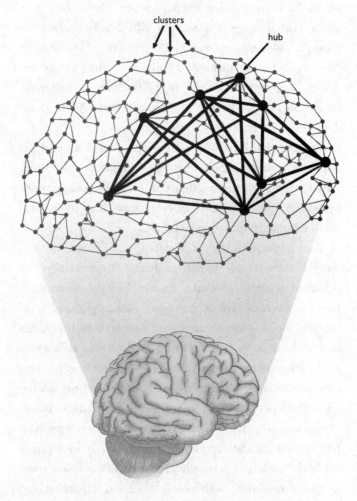

Clusters of neurons connected by hubs

world. So imagine what happens when a brain hub goes down. Hub damage is associated with depression, schizophrenia, dyslexia, chronic pain, dementia, Parkinson's disease, and other disorders. Hubs are points of vulnerability because they are points of efficiency—they make it possible to run a human brain in a human body without depleting a body budget.

You can thank natural selection for this lean and potent hub structure. Scientists speculate that over evolutionary time, neurons organized into this kind of network because it's powerful and fast yet energy-efficient and still small enough to fit inside your skull.

Your brain network is not static—it changes continuously. Some changes are extremely fast. Your brain wiring is bathed in chemicals that complete the local connections between neurons. These chemicals, such as glutamate, serotonin, and dopamine, are called neurotransmitters, and they make it easier or harder for signals to pass across synapses. They're like the airport staff —ticket agents, security screeners, ground crew—who can speed up or slow down the flow of passengers within an airport and without whom we can't travel at all. These network changes happen instantaneously and continually, even as your physical brain structure seems unchanged. In addition, some of these chemicals, such as serotonin and dopamine, can also act on *other neurotransmitters* to dial up or dial down their effects. When brain chemicals act in this way, we call them neuromodulators. They are like the weather between airports. When it's clear, planes fly quickly. When it's stormy, flights are grounded or rerouted. Neuromodulators and neurotransmitters together allow your brain's single structure to take on trillions of different patterns of activity.

Other network changes are relatively slower. Just as airports build or renovate their terminals, your brain is constantly under construction. Neurons die, and in some parts of the human brain, neurons are born. Connections become more or less numerous, and they become stronger when neurons fire together and weaker when they don't. These changes are examples of what scientists call *plasticity*, and they occur throughout your life. Anytime you learn something—a new friend's name or an interesting fact from the news—the experience becomes encoded in your wiring so you can remember it, and over time, these encodings can change that wiring.

Your network is also dynamic in another way. As neurons change conversation partners, a single neuron can take on different roles. For example, your ability to see is so intimately tied to an area of the brain called the occipital cortex that the area is routinely called the visual cortex; however, its neurons routinely carry information about hearing and touch. In fact, if you blindfold people with typical vision for a few days and teach them to read braille, neurons in their visual cortex become more devoted to the sense of touch. Remove the blindfold, and the effect disappears after twenty-four hours. Similarly, when babies are born with dense cataracts, meaning the brain receives no visual input, neurons in the visual cortex become repurposed for other senses.

Some neurons in your brain are so flexibly connected that their main job is to have many jobs. An example is one part of your famous prefrontal cortex, called the dorsomedial prefrontal cortex. This brain region is always engaged in body budgeting, but it's also regularly involved in memory, emotion, perception, decision-making, pain, moral judgments, imagination, language, empathy, and more.

Overall, no neuron has a single psychological function, though a neuron may be *more likely* to contribute to some functions than others. Even when scientists name a brain area after a function, like "visual cortex" or the "language network," the name tends to reflect the scientist's focus at the time rather than any exclusive job performed by that part of the brain. I'm not saying that every neuron can do everything, but any neuron can do *more than one* thing, just like a single airport can launch planes, sell tickets, and serve crappy food.

It's also the case that different groups of neurons can produce the same result. Try this right now: Reach for something in front of you, like your phone or a chocolate bar. Draw your hand back, and reach for it again in exactly the same way. Even a simple reaching action like this, when done more than once, can be guided by different sets of neurons. This phenomenon is called *degeneracy*.

Scientists suspect that all biological systems have degeneracy. In genetics, for example, the same eye color can be produced by different combinations of genes. Your sense of smell works by degeneracy too, and so does your immune system. Transportation systems have degeneracy as well. You can fly from London to Rome on different airlines, on different flights, on different models of airplane, in different seats, with different flight attendants. Copilots can take over for pilots. Degeneracy in the brain means that your actions and experiences can be created in multiple ways. Each time you feel afraid, for example, your brain may construct that feeling with various sets of neurons.

We've now seen how helpful it is to understand the brain as a network. This perspective captures so much of the brain's dynamic behavior—slow changes by plasticity, faster changes by neurotransmitters and neuromodulators, and the flexibility of neurons with multiple jobs.

A network organization has another advantage as well. It furnishes a brain with a special characteristic that is key to creating a human mind. This characteristic is called *complexity*. It is a brain's ability to configure itself into an *enormous number of distinct neural patterns*.

In general, a system with complexity is made of many interacting parts that collaborate and coordinate to create a multitude of patterns of activity. The world's air-travel system has complexity because its parts—the ticket agents, air traffic controllers, pilots, planes, ground crew, and so on—depend on one another to make the whole system function. The behavior of a complex system is more than the sum of its parts.

Complexity empowers a brain to act flexibly in all kinds of situations. It opens a door so we can think abstractly, have a rich, spoken language, imagine a future very different from the present, and have the creativity and innovation to construct airplanes and suspension bridges and robot vacuum cleaners. Complexity also helps us contemplate the whole world beyond our immediate surroundings, even outer space, and care about the past and the future to an extent that other animals do not. Complexity alone doesn't give us these capabilities; many other animals have complex brains too. But complexity is a critical ingredient for these capabilities, and the human brain has it in abundance.

In the brain's case, what constitutes complexity? Picture billions of neurons, each one sending signals to other specific neurons all at once, using neurotransmitters, neuromodulators, and other dynamic bells and whistles. That whole picture is one "pattern" of brain activity. Complexity means your brain can create massive numbers of different patterns by combining bits and pieces of old patterns it has made before. The result is a brain that runs its body efficiently in a world full of constantly changing situations by recalling patterns that helped in the past and generating new ones to try.

A system has higher or lower complexity depending on how much information it can manage by reconfiguring itself. The world's air-travel system is highly complex in this way. Passengers can fly almost anywhere by different combinations of flights. If a new airport opens, the system can reconfigure to accommodate it. If an airport is damaged by a tornado, travel will be disrupted for a time, but ultimately the airlines can route around the problem. A system with lower complexity, in contrast, could not reconfigure itself as well. The air-travel system would have lower complexity if any given route had just one flight plan or if all planes were forced to fly in and out of a single hub. If that hub was lost, the whole air-travel system would grind to a halt.

We can explore higher and lower complexity by considering two imaginary human brains that are less complex than yours. The first imaginary brain has about 128 billion neurons like yours does, but every neuron is connected to every other. When one neuron receives a signal to change its firing rate, all the other neurons eventually change in kind, because they are all connected. We'll call this one Meatloaf Brain because its structure is so uniform. Functionally speaking, Meatloaf Brain has lower complexity than yours because at any point in time, its 128 billion elements are effectively just a single element.

A second imaginary brain also has 128 billion neurons,

but it's carved into puzzle pieces that serve dedicated functions—seeing, hearing, smelling, tasting, touching, thinking, feeling, and so on—like the brain imagined by phrenologists in the nineteenth century. This brain is like a collection of specialized tools that work together, so we'll call it Pocketknife Brain. Pocketknife Brain has higher complexity than Meatloaf Brain but much lower complexity than your brain, because each tool adds little to the total number of patterns that Pocketknife Brain can make. A real pocketknife with, say, fourteen tools can open into about sixteen thousand possible patterns (2^{14} to be precise), and adding a fifteenth tool merely doubles the total. Your brain's neurons, however, have multiple functions that increase the number of patterns exponentially. If you had a fourteen-tool pocketknife and added one additional function to each tool—say, making the blade serve as a crude bottle opener, using the screwdriver to punch holes, and so on—the total number of patterns leaps from sixteen thousand (2^{14}) to over four million (3^{14}). In other words, when existing brain parts become more flexible, the result is *much* more complexity than we get by accumulating new parts.

Meatloaf Brain and Pocketknife Brain may have some advantages, but a brain with high complexity beats them both.

Brains of higher complexity can remember more. A brain doesn't store memories like files in a computer—it reconstructs them on demand with electricity and swirling chemicals. We call this process *remembering* but it's really *assembling*. A complex brain can assemble many more memories than either Meatloaf Brain or Pocketknife Brain could. And each time you have the same memory, your brain may have assembled it with a different collection of neurons. (That's degeneracy.)

Brains of higher complexity are also more creative. A complex brain can combine past experiences in new ways to deal with things that it has never encountered before; for example, you can climb an unfamiliar hill or staircase without tripping because you've climbed similar ones in the past. Complex brains may adjust faster to changing environments that require different body budgeting. It's one reason that humans can live successfully in so many climates and social structures. If you have to move from the equator to Northern Europe, or from a laid-back culture to one with strict rules, you'll adapt more swiftly with a complex brain in your head.

On top of that, higher complexity may make a brain more resilient to injury. If one collection of neurons stops working, other collections may serve in its place. That's one reason complex brains may be favored by natural selection. Pocketknife Brain would not have this capability; lost neurons would be more likely to mean lost functionality.

Human brains may be some of the most complex brains on Earth, but they aren't the only ones with high complexity. Intelligent behavior has emerged many times in different species with differently structured brains. Take the octopus, for example, whose complex brain is distributed throughout its body. Octopuses can solve puzzles and even dismantle their tanks in aquariums. Bird brains can be complex too. Some bird species can use simple tools and have a bit of language ability, even though their neurons are not organized into a cerebral cortex. The highly complex human brain isn't a pinnacle of evolution, remember; it's just well adapted to the environments we inhabit.

High complexity may be a prerequisite for so much that makes you human, but by itself it does not empower a human brain to make a human mind. Your Paleolithic ancestors needed more than a highly complex brain to pick up a hunk of rock and imagine a future hand axe within it. Likewise, you need more than high complexity to look at a piece of paper, a piece of metal, and a piece of plastic, which are all physically different, and treat them all as having a similar function, like serving as money. High complexity helps you climb an unfamiliar staircase, but you need more than high complexity to understand what it means for someone to climb a social ladder to gain power and influence. We also need more than high complexity to contemplate the nature of a human brain and to invent the many creative metaphors for what a brain is like, such as the triune brain, Systems 1 and 2, and mental organs. These feats of imagination require a high level of complexity packaged in a really big brain, as well as other factors that you'll learn about in the coming lessons.

A brain network is not a metaphor, as I mentioned earlier; it's the best scientific description of a brain today. It allows us to consider how one physical structure reconfigures in an instant to integrate vast amounts of information efficiently. It reveals similarities and differences between various kinds of brains by quantifying their complexity. It even helps us understand how a brain might compensate when it's damaged.

Still, I've relied on a few metaphors to explain the network. For example, the word *wiring* is a metaphor. Neurons aren't literally wired together—they're separated by the small gaps we called synapses, and chemicals complete the connections. Neurons are also not trees with branches and trunks. And your brain most likely doesn't have airports inside it.

Metaphors are wonderful for explaining complex topics in simple, familiar terms. A metaphor's simplicity, however, can become its greatest failing if people treat the metaphor as an explanation. In biology, for example, genes are sometimes described as "blueprints." If you take this metaphor literally, you might think that particular genes always have the same basic function; say, to make a specific characteristic or body part. (They don't.) Physicists sometimes say that light travels in waves, a metaphor that invites us to assume that space, like an ocean, contains some substance for those waves to move through. (It doesn't.) Metaphors provide the illusion of knowledge, so they must be used with care.

The complex network in your head may not be a metaphor, but my description here is necessarily incomplete. Your brain is more than just neurons. It includes blood vessels and various fluids that I haven't talked about. It also includes other kinds of brain cells, called glial cells, that function in ways that scientists don't fully understand yet. Your brain network may even extend, surprisingly, into your gut and intestines, where scientists have found microbes that communicate with your brain via neurotransmitters.

As scientists learn more about the brain and its interconnections, we may discover better ways to describe its structure and function. Until then, understanding the brain as a complex network allows us to ponder how a human brain creates a human mind without any need for an allegedly rational and oversize neocortex. If human brain evolution has a crowning achievement, it is the complexity of its crown.

Lesson No. 3

*Little Brains Wire Themselves
to Their World*

HAVE YOU EVER noticed that many newborn animals are more competent than newborn humans? A newborn garter snake can slither on its own almost instantly. Horses can walk shortly after birth, and an infant chimp can cling to its mother's hair. In comparison, human newborns are pretty pathetic. They can't even control their limbs. It takes weeks before they can swat their tiny hands with intent. Many animals emerge from the egg or womb with brains that are more fully wired to control their bodies, but little human brains are born under construction. They don't take on their full adult structure and function until they finish their principal wiring, a process that takes about twenty-five years.

Why did we evolve this way, to be born with our brain wiring only partially complete? No one knows for sure (though plenty of scientists have been happy to speculate). What we *can* learn is where those wiring instructions come from after birth and what advantages this arrangement affords us.

Scholars usually discuss this issue in terms of nature versus nurture—which aspects of humanity are built into our genes before birth and which ones we learn from our culture. But this distinction is illusory. We cannot attribute causes to genes alone or to the environment alone, because the two are like lovers in a fiery tango—so deeply entwined that it's unhelpful to call them separate names like *nature* and *nurture*.

To a remarkable extent, a baby's genes are guided and regulated by the surrounding environment. The brain areas that are most centrally involved in vision, for example, develop normally after birth only if a baby's retinas are regularly exposed to light. An infant's brain also learns to locate sounds in the world based on the specific shape of the baby's ear. To make matters even stranger, a baby's body requires some additional genes that sneak in from the outside world. These tiny visitors travel inside of bacteria and other critters and affect the brain in ways that scientists are only beginning to understand.

A baby's wiring instructions come not only from the physical environment but also from the social environment, from caregivers and people like you and me. When you cradle a newborn girl in your arms, you present your face to her at just the right distance to teach her brain to process and recognize faces. When you

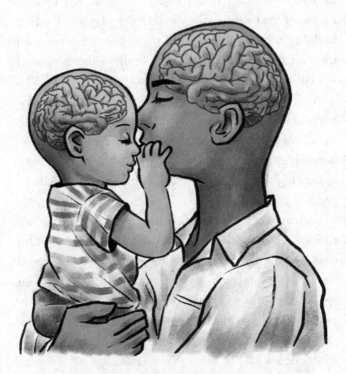

Caregivers play a critical role in wiring a baby's brain.

expose her to boxes and buildings, you're training her visual system to see edges and corners. Many other social things we do with a baby, like cuddling and talking and making eye contact in key moments, sculpt her brain in necessary and irrevocable ways. Genes play a key role in building a baby's brain wiring, and they also open the door for us to wire her newborn brain in the context of her culture.

As information travels from the world into the newborn brain, some neurons fire together more frequently than others, causing gradual brain changes that we've called plasticity. These changes nudge the infant's brain toward higher complexity via two processes we'll call *tuning* and *pruning*.

Tuning means strengthening the connections between neurons, particularly connections that are used frequently or are important for budgeting the resources of your body (water, salt, glucose, and so on). If we think again of neurons as little trees, tuning means that the branch-like dendrites become bushier. It also means that the trunk-like axon develops a thicker coating of myelin, a fatty "bark" that's like the insulation around electrical wires, which makes signals travel faster. Well-tuned connections are more efficient at carrying and processing information than poorly tuned ones and are therefore more likely to be reused in the future. This means the brain is more likely to recreate certain neural patterns that include those well-tuned connections. As neuroscientists like to say, "Neurons that fire together, wire together."

Meanwhile, less-used connections weaken and die off. This is the process of *pruning*, the neural equivalent of "If you don't use it, you lose it." Pruning is critical in a developing brain, because little humans are born with many more connections than they will ultimately use. A human embryo creates twice as many neurons as an adult brain needs, and infant neurons are quite a bit bushier than neurons in an adult brain. Unused connections are helpful at the outset. They enable a brain to tailor itself to diverse environments. But over the longer term, un-used connections are a burden, metabolically speaking— they don't contribute anything worthwhile, so it's a waste of energy for the brain to maintain them. The good news is that pruning these extra connections makes room for more learning—that is, for more useful connections to be tuned.

Tuning and pruning happen continuously and often si-multaneously, driven by the physical and social world out-side the infant's head and by the growth and activity in the infant's body. Both processes also continue through-out life. Your bushy dendrites keep sprouting new buds, and your brain tunes and prunes them. Buds that aren't tuned disappear within a couple of days.

Let's look at three examples of tuning and pruning that set newborn brains on a path to develop into typical adult brains. These examples demonstrate how our unfinished wiring completes itself in the months and years after we're born, driven by wiring instructions that arrive from the outside world.

First, consider how you manage your body budget. When you're hungry, you can open the fridge. When you're tired, you can go to bed. When you're cold, you can put on a coat. When you're agitated, you can take deep breaths to calm your nerves. Babies can't do any of these things by themselves. They can't even burp without help.

That's where caregivers come in. They regulate the baby's physical environment and therefore her body budget by feeding her, setting sleep times (or trying to!), and wrapping her in blankets and cuddles. These actions help the baby's brain maintain its body budget, so her internal systems operate efficiently and she stays alive and healthy.

If caregivers do these activities effectively, the baby's brain is free to tune and prune itself to perform healthy body budgeting. Little by little, the caregivers' roles diminish as the infant's brain becomes more capable of controlling her body, enabling her to fall asleep without being held or to stuff a bit of banana into her mouth without smearing it on her face. It may take years before the little brain can put on a sweater by herself or make her own breakfast, but eventually she'll have primary responsibility for her own body budget.

Little brains are also wired by what caregivers *don't* do. If you don't let a baby fall asleep on her own and instead rock her to sleep every night, her brain might not learn how to fall asleep without help. When an infant is crying for a long time and you don't check in regularly, her brain may learn that the world is unreliable and unsafe while her body budget goes untended.

Things change once she's a toddler, however. Her toddler brain has to learn to calm her body after a tantrum and, eventually, to body-budget without a tantrum in the first place. When my daughter was little, I found it helpful to give her space so her brain could learn to soothe her body. In general, toddlers learn to tend their own body budgets better when their caregivers create learning opportunities for them instead of hovering and taking care of their every need. A big challenge of parenting is knowing when to step in and when to step back.

Our second example of tuning and pruning concerns how you learn to pay attention. Have you ever been in a crowd, not really attending to the conversations sprouting around you, and then someone speaks your name and you immediately orient to it? (Scientists call this the "cocktail party effect.") Your adult brain can effortlessly focus on one thing and ignore others, similar to a spotlight in the darkness. That's because your brain network contains smaller communities of neurons whose main job is to focus on certain details as important and ignore other details as irrelevant. Your brain focuses its spotlight of attention continually and automatically, and often you're unaware that it's happening.

We do need help sometimes to focus our spotlight—
that's why noise-canceling headphones sell so well. But
the newborn brain doesn't have a spotlight. It has more
of a lantern, illuminating a wide area in its physical environment. Newborn brains don't know what's important
and what's not, so they cannot focus as adults do. They
still lack the wiring that narrows their lantern into a spotlight.

Again, the missing ingredient comes from caregivers
in the social world. They constantly guide the baby's attention to things of interest. A mother picks up a toy dog
and looks at it. She looks at her little boy, then back at
the dog, guiding the baby's gaze. She turns to her son and
says with intent, "What a cute little doggie," in a singsong
tone. The mother's speech and the back-and-forth switch-
ing of gaze, which scientists call sharing attention, alert
the baby that the toy dog is significant—that is, the toy
could affect his body budget, so he should care about and
learn about the dog.

Little by little, sharing attention teaches an infant
which parts of the environment matter and which parts
don't. The infant brain is then able to construct its own
environment of what is relevant to its body budget and
what can be ignored. Scientists call this environment a
niche. Every animal has a niche, and it creates that niche
as it senses the world, makes worthwhile movements, and
regulates its body budget. Adult humans have a gigantic
niche, perhaps the largest of any creature. Your niche ex-

tends far beyond your immediate surroundings to include events around the world, past, present, and future.

After months of practice sharing attention with caregivers, an infant will learn to elicit shared attention from them. He will look at them as a way of asking whether something is in his niche and what it might mean for his body budget. In this manner, the infant learns to focus attention even more effectively on things that matter.

Our third example of tuning and pruning is how your senses develop. In the first few months of life, babies are bathed in all kinds of sounds, including the sound of people speaking. Newborns, with their lantern of attention, take in all the sounds around them. When tested in a lab, newborns can distinguish a wide range of language sounds, including those that they don't hear very often. But over time, tuning and pruning will wire the baby's brain based on the vocal sounds he hears more regularly. Sounds that are frequent cause certain neural connections to be tuned, and the baby's brain starts to treat those sounds as part of its niche. Sounds that are rare are treated as noise to be ignored, and eventually, related neural connections fall out of use and are pruned away.

Scientists think this sort of pruning may be one reason why children have an easier time learning languages than adults do. Different spoken languages use different sets of sounds. For example, Greek and Spanish have a handful of vowel sounds, while Danish has twenty or more (depending on how they're counted). If people interacted with you in multiple languages when you were a baby, then your brain was likely tuned and pruned to hear and distinguish the sounds in those languages. If you heard only one language as a baby, you'd need to relearn the ability to hear and distinguish sounds outside your language, which is hard.

This process works similarly for seeing faces. When you were a baby, you learned to recognize the people around you. Your infant brain was tuned and pruned to detect fine differences in their faces so you could tell them apart. But there's a catch — people tend to live around others of the same ethnicity, so babies are often not exposed to a wide array of facial features. That means the baby's brain does not tune itself to detect those different features. Scientists think this is one reason why it can be harder for you to remember the faces of people of an ethnicity different from your own or to tell one face from another. Fortunately, you can quickly retune your brain and restore this ability by looking at lots of diverse faces; it's much easier than retuning to the sounds of a foreign language.

These examples of hearing language and seeing faces focus on a single sense, but you live in a multisensory world. For example, when you kiss someone, you are enveloped in a unified experience that combines the sight of a face, the sound of breathing, the feel, taste, and scent of luscious lips, and the racing of your heart. Your brain assembles these sensations into a cohesive whole. Scientists call this process *sensory integration*.

Sensory integration itself is tuned and pruned as babies grow. A newborn at first can't recognize his mother by her face, because he hasn't learned what a face is, and his visual system isn't fully formed. He might know a bit about how his mother sounds, and he can smell her breast milk. If you put a newborn on his mother's belly, he will wriggle up to her breast by following the aroma. Soon, he learns to recognize his mother by different combinations of all his senses together. His little brain absorbs each pattern of sight, smell, sound, touch, and taste, plus sensations from inside his body, and learns its meaning: the person who regulates his body budget is here. Sensory integration conjures his first feeling of trust. It's part of the neural foundation for attachment.

Our three examples of tuning and pruning demonstrate how the social world profoundly shapes the physical reality of the brain's wiring. Who knew that caregivers were such effective electricians?

This arrangement comes with a risk, however. Little brains *require* a social world in order to develop typically. You've already learned that babies need certain physical inputs, such as photons of light bombarding their retinas, or their brains will never develop normal vision. It turns out that they also need social inputs from other humans who guide their attention, speak or sing to them, and cuddle them at key moments. If these needs aren't met, things can go terribly wrong.

I wish we didn't know what happens when a baby's

brain receives too little social input. Nobody should ever deprive babies of what they need to thrive. But unfortunately, we do know some distressing details because of a tragic historical event.

In the 1960s, the Communist government of Romania outlawed most contraception and abortion. The president, Nicolae Ceauşescu, wanted to expand the population to become more of an economic power and, therefore, a world power. This new law produced a huge increase in births, more children than many families could afford. As a consequence, hundreds of thousands of children were sent to live in orphanages. Many were appallingly mistreated. The children who are most relevant to our lesson here are the ones whose social needs went unmet.

In some orphanages, babies were warehoused in rows of cribs, with little stimulation or social interaction. Nurses or caregivers would come in and feed them, change them, and put them back in the cribs. That was about it. Nobody cuddled these babies. No one played with them. No one conversed with or sang to them, or shared attention. They were ignored.

As a consequence of this social neglect, the Romanian orphans grew up intellectually impaired. They had problems learning language. They had difficulty concentrating and resisting distractions, probably because nobody had shared attention with them, so their brains never developed the wiring for an effective spotlight. They also had trouble controlling themselves. Alongside the children's mental and behavioral issues, their bodies were stunted, most likely because they grew up without caregivers to keep their body budgets solvent. This meant their brains never learned to budget effectively. A little brain wires itself to its environment, and when that environment is missing key elements for healthy body budgeting, critical brain wiring can be pruned away.

These aftereffects are consistent with what scientists know about other babies raised in severely socially impoverished conditions. Their brains develop smaller than average. Key brain regions are smaller too, and important areas of the cerebral cortex have fewer connections. If such children are moved to traditional foster homes in the first few years of life, some of these effects are reversible. Similar risks can arise for any kids reared in institutions without attentive, consistent caregivers, whether these institutions are orphanages, refugee camps, or immigrant detention centers.

When children are persistently neglected, in all likelihood they'll suffer ill effects eventually. The impact might not be immediate and dramatic, as in the Romanian orphanages, but can be gradual and subtle as important wiring goes unused and is steadily pruned away. The fallout may build up over time like a slow drip from a water pipe that eventually bores a hole through your floorboards. For example, a neglected little brain in a socially impoverished environment may wire itself to manage its own body budget alone, without the social support from caregivers and the wiring instructions they provide through their actions. This nontypical wiring imposes a pernicious burden on the body budget that accumulates over years, raising the odds of serious health problems later, such as heart disease, diabetes, and mood disorders like depression, all of which have metabolic underpinnings.

To be clear, I'm not saying that we have to keep our little darlings free of stress or their brains and bodies will break. I'm saying that *persistent* neglect, over a long time with no relief, is almost always harmful to a little brain. The scientific evidence is clear on this point. You can't just feed and water babies and expect their brains to grow normally. You must also meet their social needs with eye contact and language and touch. If these needs go unmet, the seeds of illness may be planted very, very early.

We see similar consequences when little brains develop in poverty. Research shows that early and long exposure to poverty is bad for the developing brain. Inadequate nutrition, interrupted sleep due to street noise, poor temperature regulation due to lack of heat or ventilation, and other circumstances of poverty may alter the development of the front of the cerebral cortex, namely the prefrontal cortex. This brain area is involved in a range of critical functions, including attention, language, and body budgeting. Scientists are still studying the ways that poverty affects brain development, but we do know that it's linked to poorer performance in school and fewer years of education. These burdens ultimately increase a child's risk of living in poverty when he grows up and has children of his own. It wouldn't surprise me if this vicious cycle reinforces negative stereotypes about people who live in poverty. Society is quick to blame genes when poverty endures across generations for a group of people. But it's plausible that those little brains are being molded *by* poverty.

Some kids are fortunate enough to be naturally resilient to the insidious effects of adversity and poverty. But on average, adversity and poverty are afflictions from which little brains struggle to recover. What's truly frustrating is that this tragedy is *preventable*. (Pardon me while I take off my scientist's hat for a moment.) Politicians have dragged their feet for decades about lifting children out of poverty. So let's set the politics aside and frame the issue in simple financial terms: Childhood poverty is a colossal waste of human opportunity. Recent estimates suggest that it's far cheaper to eradicate poverty than to deal with its effects decades later. More school districts can offer free meal programs to students in need. Cities can set up noise ordinances for poor neighborhoods. These sorts of steps are not merely about quality of life. They create the conditions for healthy brain development, so all children can become the workers, citizens, and innovators of the next generation.

Given the powerful impact of neglect and poverty on a little brain, it's tempting to ask how evolution got our species into this precarious situation in the first place. It's risky for a baby's brain wiring to depend so critically on social and physical input in order to develop typically. We humans must gain some advantage to offset the risks of developing this way. So what is it?

We can't know for sure, but here is my guess based on evidence from evolutionary biology and anthropology: This arrangement helps our cultural and social knowledge flow efficiently from generation to generation. Each little brain becomes optimized for its particular environment, the one it developed in. Caregivers curate a baby's physical and social niche, and the baby's brain learns that niche. When the baby grows up, he perpetuates that niche by passing his culture to the next generation through his words and actions, wiring their brains in turn. This process, called cultural inheritance, is efficient and frugal because evolution doesn't have to encode all our wiring instructions in genes. It off-loads much of the job to the world around us, including the other humans in that world. We unknowingly wire the knowledge of our culture into our offspring after birth, for better or worse.

When it comes to the brain, simple distinctions like nature versus nurture are alluring but not realistic. We have the kind of nature that *requires* nurture. Your genes require a physical and social environment—a niche filled with other humans who shared your infant gaze, spoke to you with intent, set your sleep schedule, and controlled your body temperature—in order to produce a finished brain.

We all know that it matters how we treat our children, but it matters more than we knew even a few decades ago. When you're awake at four a.m. trying to console your shrieking little angel, or when he calmly drops his Cheerios onto the floor for the ninety-third time, you are guiding his tuning and pruning whether you know it or not. Little brains wire themselves to their world. It's up to us to create that world—including a social world rich with wiring instructions—to grow those brains healthy and whole.

Lesson No. 4

*Your Brain Predicts (Almost)
Everything You Do*

A FEW YEARS AGO, I received an e-mail from a man who served in the Rhodesian army in southern Africa in the 1970s, before the end of apartheid. He'd been drafted against his will, handed a uniform and a rifle, and ordered to hunt down guerrilla fighters. To make matters worse, before the draft, he'd been an advocate for the same guerrillas that he was now required to treat as the enemy.

He was deep in the forest one morning, conducting practice exercises with his small squad of soldiers, when he detected movement ahead of him. With a pounding heart, he saw a long line of guerrilla fighters dressed in camouflage and carrying machine guns. Instinctively, he raised his rifle, flipped off the safety catch, squinted down the barrel, and aimed at the leader, who was carrying an AK-47 assault rifle.

Suddenly, he felt a hand on his shoulder. "Don't shoot," whispered his buddy behind him. "It's just a boy." He slowly lowered his rifle, looked again at the scene, and was astonished by what he now saw: a boy, perhaps ten years old, leading a long line of cows. And the dreaded AK-47? It was a simple herding stick.

For years afterward, this man struggled to understand the unsettling episode. How had he managed to mis-see what was right in front of his eyes and nearly kill a child? What was wrong with his brain?

As it turns out, nothing was wrong with his brain. It was working exactly as it should have.

Scientists used to believe that the brain's visual system operated sort of like a camera, detecting the visual information "out there" in the world and constructing a photograph-like image in the mind. Today we know better. Your view of the world is no photograph. It's a construction of your brain that is so fluid and so convincing that it appears to be accurate. But sometimes it's not.

To understand why it can be perfectly normal to see a grown guerrilla fighter with a rifle when you're looking at a ten-year-old boy with a stick, let's consider the situation from the brain's point of view.

From the moment you're born to the moment you draw your last breath, your brain is stuck in a dark, silent box called your skull. Day in and day out, it continually receives sense data from the outside world via your eyes, ears, nose, and other sensory organs. This data does not arrive in the form of the meaningful sights, smells, sounds, and other sensations that most of us experience. It's just a barrage of light waves, chemicals, and changes in air pressure with no inherent significance.

Faced with these ambiguous scraps of sense data, your brain must somehow figure out what to do next. Remember, your brain's most important job is to control your body so you stay alive and well. Your brain must somehow make meaning from the onslaught of sense data it's receiving so you don't fall down a staircase or become lunch for some wild beast.

How does your brain decipher the sense data so it knows how to proceed? If it used only the ambiguous information that is immediately present, then you'd be swimming in a sea of uncertainty, flailing around until you figured out the best response. Luckily, your brain has an additional source of information at its disposal: memory. Your brain can draw on your lifetime of past experiences—things that have happened to you personally and things that you've learned about from friends, teachers, books, videos, and other sources. In the blink of an eye, your brain reconstructs bits and pieces of past experience as your neurons pass electrochemical information back and forth in an ever-shifting, complex network. Your brain assembles these bits into memories to infer the meaning of the sense data and guess what to do about it.

Your past experiences include not only what happened in the world around you but also what happened inside your body. Was your heart beating quickly? Were you breathing heavily? Your brain asks itself in every moment, figuratively speaking, *The last time I encountered a similar situation, when my body was in a similar state, what did I do next?* The answer need not be a perfect match for your situation, just something close enough to give your brain an appropriate plan of action that helps you survive and even thrive.

This explains how the brain plans your body's next action. How does your brain also conjure high-fidelity experiences, like guerrilla fighters in the forest, out of scraps of raw data from the outside world? How does it create feelings of terror from a thundering heart? Once again, your brain recreates the past from memory by asking itself, *The last time I encountered a similar situation, when my body was in a similar state and was preparing this particular action, what did I* see *next? What did I* feel *next?* The answer becomes your experience. In other words, your brain combines information from outside *and inside* your head to produce everything you see, hear, smell, taste, and feel.

Here's a quick demonstration that your memory is a critical ingredient in what you see. Take a look at the three line drawings on page 68.

Excerpted from The Ultimate Droodles Compendium by Roger Price.

What do you see?

Inside your skull, without your awareness, billions of your neurons are trying to give these lines and blobs meaning. Your brain is searching through a lifetime of past experiences, issuing thousands of guesses at once, weighing probabilities, trying to answer the question *What are these wavelengths of light most like?* And it's all happening faster than you can snap your fingers.

So what do you see? A bunch of black lines and a couple of blobs? Let's see what happens when we give your brain some more information. Turn to pages 153–54 of the appendix, read the entry for *line drawings*, and then come back and look at the drawings again.

You should now see familiar objects instead of lines and blobs. Your brain is assembling memories from bits and pieces of past experiences to go beyond the visual data in front of you and make meaning. In the process, your brain is literally changing the firing of its own neurons. Objects that you might never have seen before now leap from the page. The lines and blobs haven't changed —you have.

Artwork, particularly abstract art, is made possible because the human brain constructs what it experiences. When you view a cubist painting by Picasso and see recognizable human figures, that happens only because you have memories of human figures that help your brain make sense of the abstract elements. The painter Marcel Duchamp once said that an artist does only 50 percent of the work in creating art. The remaining 50 percent is in the viewer's brain. (Some artists and philosophers call the second half "the beholder's share.")

Your brain actively constructs your experiences. Every morning, you wake up and experience a world around you full of sensations. You might feel the bedsheets against your skin. Maybe you hear sounds that woke you, like an alarm buzzing or birds chirping or your spouse snoring. Perhaps you smell coffee brewing. These sensations seem to sail right into your head as if your eyes, nose, mouth, ears, and skin were transparent windows on the world. But you don't sense with your sensory organs. You sense with your brain.

What you see is some combination of what's out there in the world and what's constructed by your brain. What you hear is also some combination of what's out there and what's in your brain, and likewise for your other senses.

In much the same way, your brain also constructs what you feel inside your body. Your aches and jitters and other inner sensations are some combination of what's going on in your brain and what's actually happening within your lungs and heart and gut and muscles and so on. Your brain also adds information from your past experiences to guess what those sensations mean. For instance, when people haven't slept enough and are fatigued or low energy, they may feel hungry (because they've been hungry before when their energy was low) and may think that a quick snack will boost their energy. In fact, they're just tired from lack of sleep. This constructed experience of hunger may be one reason why people gain unwanted weight.

Now we can unravel why our soldier friend saw guerrilla fighters instead of a shepherd boy with cows. His brain asked, *Based on what I know about this war, and given that I am deep in the woods with my comrades, gripping a rifle, heart pounding, and there are moving figures ahead, and maybe something pointy, what am I likely to see next?* And the result was *Guerrilla fighters.* In this situation, the stuff inside and outside his head didn't match, and the inside stuff prevailed.

Most of the time when you look at cows, you see cows. But you've almost certainly had an experience like the soldier's, where the information inside your head triumphs over the data from the outside world. Have you ever seen a friend's face in a crowd, but when you looked again, you realized it was a different person? Have you ever felt your cell phone vibrate in your pocket when it didn't? Have you ever had a song playing in your head that you couldn't get rid of? Neuroscientists like to say that your day-to-day experience is a carefully controlled hallucination, constrained by the world and your body but ultimately constructed by your brain. It's not the kind of hallucination that sends you to the hospital. It's an everyday kind of hallucination that creates all your experiences and guides all your actions. It's the normal way that your brain gives meaning to your sense data, and you're almost always unaware that it's happening.

I realize that this description defies common sense, but wait: there's more. This whole constructive process happens *predictively*. Scientists are now fairly certain that your brain actually begins to sense the moment-to-moment changes in the world around you *before* those light waves, chemicals, and other sense data hit your brain. The same is true for moment-to-moment changes in your body — your brain begins to sense them before the relevant data arrives from your organs, hormones, and various bodily systems. You don't experience your senses this way, but it's how your brain navigates the world and controls your body.

But don't take my word for it. Instead, think of the last time you were thirsty and drank a glass of water. Within seconds after draining the last drops, you probably felt less thirsty. This event might seem ordinary, but water actually takes about twenty minutes to reach your bloodstream. Water can't possibly quench your thirst in a few seconds. So what relieved your thirst? Prediction. As your brain plans and executes the actions that allow you to drink and swallow, it simultaneously anticipates the sensory consequences of gulping water, causing you to feel less thirsty long before the water has any direct effect on your blood.

Predictions transform flashes of light into the objects you see. They turn changes in air pressure into recognizable sounds, and traces of chemicals into smells and tastes. Predictions let you read the squiggles on this page and understand them as letters and words and ideas. They're also the reason why it feels unsatisfying when a sentence is missing its final.

Scientists have had hints for more than a century that brains are predicting organs, though we didn't decipher those hints until recently. You might have heard of Ivan Pavlov, the nineteenth-century physiologist who famously taught his dogs to salivate upon hearing a sound (usually described as a bell, but it was really a ticking metronome). Pavlov played that sound right before his dogs ate each meal, and eventually the dogs salivated when they heard the sound even when they weren't fed. Pavlov won a Nobel Prize for discovering this effect, which became known as Pavlovian or classical conditioning, but he didn't realize that he was discovering how brains predict. His dogs were not reacting to the sound by drooling. Their brains were predicting the experience of eating food and preparing their bodies in advance to consume it.

You can try a similar experiment right now. Picture your favorite food in your mind. (For me, it's a morsel of dark chocolate with sea salt.) Imagine its smell, its taste, and how it feels in your mouth. Are you salivating yet? I am, just writing this, and no metronome is required. If neuroscientists were scanning my brain right now, they might see increased activity in regions that are important for the sense of taste and smell and in regions that control salivation.

If this demonstration made you smell or taste your favorite food or made your mouth water even a bit, then you successfully changed the firing of your own neurons in exactly the same way that automatic predictions do. This process is similar to what happened when you looked at the three drawings earlier. In both cases, I used deliberate, contrived examples to reveal what your brain does naturally and automatically.

In a very real sense, predictions are just your brain having a conversation with itself. A bunch of neurons make their best guess about what will happen in the immediate future based on whatever combination of past and present that your brain is currently conjuring. Those neurons then announce that guess to neurons in other brain areas, changing their firing. Meanwhile, sense data from the world and your body injects itself into the conversation, confirming (or not) the prediction that you'll experience as your reality.

In actuality, your brain's predictive process is not quite so linear. Usually your brain has several ways to deal with a given situation, and it creates a flurry of predictions and estimates probabilities for each one. Is that rustling sound in the forest due to the wind, an animal, an enemy fighter, or a shepherd? Is that long, brown shape a branch, a staff, or a rifle? Ultimately, in each moment, some prediction is the winner. Often, it's the prediction that best matches the incoming sense data, but not always. Either way, the winning prediction becomes your action and your sensory experience.

So, your brain issues predictions and checks them against the sense data coming from the world and your body. What happens next still astounds me, even as a neuroscientist. If your brain has predicted well, then your neurons are *already firing* in a pattern that matches the incoming sense data. That means this sense data itself has no further use beyond confirming your brain's predictions. What you see, hear, smell, and taste in the world and feel in your body in that moment are *completely constructed in your head*. By prediction, your brain has efficiently prepared you to act.

Here's what I mean. Suppose when the soldier's brain predicted a line of guerrilla fighters up ahead, the fighters were actually there. From his brain's perspective, the real fighters confirmed the prediction, because his brain had already constructed the sights, sounds, and smells of the fighters, adjusted his body budget, and prepared his body to act. In this case, his predictions prepared him to raise his rifle and shoot.

But in the real story, the soldier's brain made the wrong prediction. It predicted a posse of guerrilla fighters with machine guns when really he faced a shepherd boy with a herding staff and a bunch of cattle. In that situation, his brain had two options. One option was to incorporate the sense data from the outside world, update his predictions, and construct a new, corrected experience of a boy and his cows. This new prediction would seed the soldier's brain and improve prediction next time. Scientists have a fancy name for this option. We call it "learning."

The soldier's brain chose the other option, however; his brain stuck with its prediction in spite of the sense data from the world. This can happen for many reasons, one being that his brain predicted his life was on the line. Brains aren't wired for accuracy. They're wired to keep us alive.

When your predicting brain is right, it creates your reality. When it's wrong, it still creates your reality, and hopefully it learns from its mistakes. It's fortunate that the soldier's friend tapped him on the shoulder, prompting him to look again and allowing his brain to launch new predictions.

Now here's the final nail in the coffin of common sense: All this predicting happens *backward* from the way we experience it. You and I seem to sense first and act second. You see an enemy and then raise your rifle. But in your brain, sensing actually comes second. Your brain is wired to prepare for action first, like moving your index finger onto a trigger and making body-budgeting changes to support that movement. It's also wired to route these predictions to your sensory systems, which predict the feeling of cold steel on your fingertip and your racing heart beat. In the case of our soldier friend, his brain heard rustling leaves, moved his hands on the gun, and guided itself to see enemies that weren't present.

Yes, your brain is wired to initiate your actions *before* you're aware of them. That is kind of a big deal. After all, in everyday life, you do many things by choice, right? At least it seems that way. For example, you chose to open this book and read these words. But the brain is a predicting organ. It launches your next set of actions based on your past experience and current situation, and it does so outside of your awareness. In other words, your actions are under the control of your memory and your environment. Does this mean you have no free will? Who's responsible for your actions?

Philosophers and other scholars have debated the existence of free will pretty much since the invention of philosophy. It's not likely that we will settle that debate here. Nevertheless, we can highlight a piece of the puzzle that is often ignored.

Think about the last time you acted on autopilot. Maybe you bit your nails. Maybe your brain-to-mouth connection was too well oiled and you muttered something regrettable to a friend. Maybe you looked away from an engaging movie and discovered that you'd downed an entire jumbo bag of red Twizzlers. In these moments, your brain employed its predictive powers to launch your actions, and you had no feeling of agency. Could you have exercised more control and changed your behavior in the moment?

Maybe, but it would have been difficult. Were you responsible for these actions? More than you might think.

The predictions that initiate your actions don't appear out of nowhere. If you hadn't chomped on your nails as a kid, you probably wouldn't bite them now. If you'd never learned the regrettable words you tossed at your friend, you couldn't say them now. If you'd never developed a taste for licorice . . . you get the idea. Your brain predicts and prepares your actions using your past experiences. If you could magically reach back in time and change your past, your brain would predict differently today, and you might act differently and experience the world differently as a result.

It's impossible to change your past, but right now, with some effort, you can change how your brain will predict in the future. You can invest a little time and energy to learn new ideas. You can curate new experiences. You can try new activities. Everything you learn today seeds your brain to predict differently tomorrow.

Here's an example. All of us have had a nervous feeling before a test, but for some people, this anxiety is crippling. Based on their past experiences of taking tests, their brains predict and launch a hammering heartbeat and sweaty hands and they're unable to complete the test. If this happens enough, they fail courses or even drop out of school. But here's the thing: a hammering heartbeat is not necessarily anxiety. Research shows that students can learn to experience their physical sensations not as anxiety but as energized determination, and when they do, they perform better on tests. That determination seeds their brains to predict differently in the future so they can get their butterflies flying in formation. If they practice this skill enough, they can pass a test, perhaps pass their courses, and even graduate, which has a huge impact on their future earning potential.

It's also possible to change predictions to cultivate empathy for other people and act differently in the future. An organization called Seeds of Peace tries to change predictions by bringing together young people from cultures that are in serious conflict, like Palestinians and Israelis, and Indians and Pakistanis. The teens participate in activities like soccer, canoeing, and leadership training, and they can talk about the animosity between their cultures in a supportive environment. By creating new experiences, these teens are changing their future predictions in the hopes of building bridges between the cultures and, ultimately, creating a more peaceful world.

You can try something similar on a smaller scale. Today, many of us feel like we live in a highly polarized world, where people with opposing opinions cannot even be civil to each other. If you want things to be different, I offer you a challenge. Pick a controversial political issue that you feel strongly about. In the United States, that might be abortion, guns, religion, the police, climate change, reparations for slavery, or perhaps a local issue that's important to you. Spend five minutes per day deliberately considering the issue from the perspective of those you disagree with, not to have an argument with them in your head, but to understand how someone who's just as smart as you can believe the opposite of what you do.

I'm not asking you to change your mind. I'm also not saying this challenge is easy. It requires a withdrawal from your body budget, and it might feel pretty unpleasant or even pointless. But when you try, really try, to embody someone else's point of view, you can change your future predictions about the people who hold those different views. If you can honestly say, "I absolutely disagree with those people, but I can understand why they believe what they do," you're one step closer to a less polarized world. This is not magical liberal academic rubbish. It's a strategy that comes from basic science about your predicting brain.

Everyone who's ever learned a skill, whether it's driving a car or tying a shoe, knows that things that require effort today become automatic tomorrow with enough practice. They're automatic because your brain has tuned and pruned itself to make different predictions that launch different actions. As a consequence, you experience yourself and the world around you differently. That is a form of free will, or at least something we can arguably call free will. We can choose what we expose ourselves to.

My point here is that you might not be able to change your behavior in the heat of the moment, but there's a good chance you can change your predictions *before* the heat of the moment. With practice, you can make some automatic behaviors more likely than others and have more control over your future actions and experiences than you might think.

I don't know about you, but I find this message hopeful, even though, as you might suspect, this extra bit of control comes with some fine print. More control also means more responsibility. If your brain doesn't merely react to the world but actively predicts the world and even sculpts its own wiring, then who bears responsibility when you behave badly? You do.

Now, when I say *responsibility*, I'm not saying people are to blame for the tragedies in their lives or the hardships they experience as a result. We can't choose everything that we're exposed to. I'm also not saying that people with depression, anxiety, or other serious illnesses are to blame for their suffering. I'm saying something else: Sometimes we're responsible for things not because they're our fault, but because we're the only ones who can change them.

When you were a child, your caregivers tended the environment that wired your brain. They created your niche. You didn't choose that niche — you were a baby. So you're not responsible for your early wiring. If you grew up around people who, say, were very similar to one another, wearing the same types of clothing, agreeing on certain beliefs, practicing the same religion, or having a narrow range of skin tones or body shapes, these sorts of similarities tuned and pruned your brain to predict what people are like. Your developing brain was handed a trajectory.

Things are different after you grow up. You can hang out with all kinds of people. You can challenge the beliefs that you were swaddled in as a child. You can change your own niche. Your actions today become your brain's predictions for tomorrow, and those predictions automatically drive your future actions. Therefore, you have some freedom to hone your predictions in new directions, and you have some responsibility for the results. Not everyone has broad choices about what they can hone, but everyone has *some* choice.

As the owner of a predicting brain, you have more control over your actions and experiences than you might think and more responsibility than you might want. But if you embrace this responsibility, think about the possibilities. What might your life be like? What kind of person might you become?

Lesson No. 5

Your Brain Secretly Works with Other Brains

WE HUMANS ARE a social species. We live in groups. We take care of one another. We build civilizations. Our ability to cooperate has been a major adaptive advantage. It has allowed us to colonize virtually every habitat on Earth and survive and thrive in more climates than any other animal, except maybe bacteria.

Part of being a social species, it turns out, is that we regulate one another's body budgets—the ways in which our brains manage the bodily resources we use every day. You've already learned how caregivers help their babies' brains to budget these resources efficiently (or badly, in the case of the Romanian orphans) as those little brains wire themselves to their world. Well, the mutual body budgeting and rewiring continue long after those little brains are grown. For your whole life, outside of your awareness, you make deposits of a sort into other people's body budgets, as well as withdrawals, and others do the same for you. This ongoing undercover operation has pros and cons with profound implications for how we live our lives.

How do the people around you influence your body budget and rewire your adult brain? Remember that your brain changes its own wiring after new experiences, a process called plasticity. Microscopic parts of your neurons change gradually every day through tuning and pruning. For example, branch-like dendrites become bushier, and their associated neural connections become more efficient. This remodeling job requires energy from your body budget, so your predicting brain needs a good reason to splurge. And a great reason is that the connections are used frequently to deal with the people around you. Little by little, your brain becomes tuned and pruned as you interact with others.

Some brains are more attentive to the people around them, and others less so, but everybody has somebody. (Even psychopaths are dependent on other people, just in a really unfortunate way.) Ultimately, your family, friends, neighbors, and even strangers contribute to your brain's structure and function and help your brain keep your body humming along.

This co-regulation has measurable effects. Changes in one person's body often prompt changes in another person's body, whether the two are romantically involved, just friends, or strangers meeting for the first time. When you're with someone you care about, your breathing can synchronize, as can the beating of your hearts, whether you're in casual conversation or a heated argument. This sort of physical connection happens between infants and their caregivers, between therapists and their clients, and among people taking a yoga class or singing in a choir together. We often mirror each other's movements in a dance that neither of us is aware of and that is choreographed by our brains. One of us leads, the other follows, and sometimes we switch. In contrast, when we don't like or trust each other, our brains are like dance partners who step on each other's toes.

We also adjust each other's body budgets by our actions. If you raise your voice, or even your eyebrow, you can affect what goes on inside other people's bodies, such as their heart rate or the chemicals carried in their bloodstream. If your loved one is in pain, you can lessen her suffering merely by holding her hand.

Being a social species has all sorts of advantages for us *Homo sapiens*. One advantage is that we live longer if we have close, supportive relationships with other people. It may seem obvious that loving relationships are good for us, but studies show that the benefits go beyond what common sense would suggest. If you and your partner feel that your relationship is intimate and caring, that you're responsive to each other's needs, and that life seems easy and enjoyable when you're together, both of you are less likely to get sick. If you're already sick with a serious illness, such as cancer or heart disease, you're more likely to get better. These studies were conducted on married couples, but the results appear to hold for close friendships too, and even for pet owners.

Another advantage of being a social species is that we do better at our jobs when we work with peers and managers whom we trust. Some employers intentionally foster that trust and reap the benefits. For example, some companies provide free meals to their workers, not just as a tasty perk but also to encourage employees to socialize and brainstorm together. Some offices also contain plenty of impromptu workspaces so employees can collaborate away from their desks. When people work in an environment where they can learn to trust one another, they'll have less burden on their body budgets, saving resources that can be invested in new ideas.

In general, being a social species is good for us, but there are also disadvantages. We may be healthier and live longer if we have close relationships, but we also get sick and die earlier when we persistently feel lonely —possibly years earlier, based on the data. Without someone else helping to regulate our body budgets, we bear an extra burden. Have you ever lost someone close to you through a breakup or a death and felt like you'd lost a part of yourself? That's because you did. You lost a source of keeping your bodily systems in balance. The poet Alfred, Lord Tennyson, famously wrote, "'Tis better to have loved and lost than never to have loved at all."

In neuroscience terms, a breakup might make you feel like you're dying, but constant loneliness is likely to hasten your death. This is one argument for why solitary confinement in jail—enforced loneliness—is like capital punishment in slow motion.

A surprising disadvantage of shared body budgeting is that it has an impact on empathy. When you have empathy for other people, your brain predicts what they'll think and feel and do. The more familiar the other people are to you, the more efficiently your brain predicts their inner struggles. The whole process feels obvious and natural, as if you were reading another person's mind. But there's a catch—when people are less familiar to you, it can be harder to empathize. You might have to learn more about the person, an extra effort that translates into more withdrawals from your body budget, which can feel unpleasant. This may be one reason why people sometimes fail to empathize with those who look different or believe different things than they do and why it can feel uncomfortable to try. It's metabolically costly for a brain to deal with things that are hard to predict. No wonder people create so-called echo chambers, surrounding themselves with news and views that reinforce what they already believe—it reduces the metabolic cost and unpleasantness of learning something new. Unfortunately, it also reduces the odds of learning something that might change a person's mind.

Besides humans, many other creatures regulate each

other's body budgets. Ants, bees, and other insects do this using chemicals such as pheromones. Mammals like rats and mice use chemicals to communicate by smell, and they add vocal sounds and touch. Primates like monkeys and chimpanzees also use vision to regulate each other's nervous systems. Humans are unique in the animal kingdom, however, because we also regulate each other with *words*. A kind word may calm you, as when a friend gives you a compliment at the end of a hard day. A hateful word from a bully may cause your brain to predict threat and flood your bloodstream with hormones, squandering precious resources from your body budget.

The power of words over your biology can span great distances. Right now, I can text the words *I love you* from the United States to my close friend in Belgium, and even though she cannot hear my voice or see my face, I will change her heart rate, her breathing, and her metabolism. Or someone could text something ambiguous to you like *Is your door locked?* and odds are that it would affect your nervous system in an unpleasant way.

Your nervous system can be perturbed not only across distances, but also across the centuries. If you've ever taken comfort from ancient texts such as the Bible or the Koran, you've received body-budgeting assistance from people long gone. Books, videos, and podcasts can warm you or give you the chills. These effects might not last long, but research shows that we all can tweak one another's nervous systems quickly with mere words in very physical ways that go beyond what you might suspect.

In my research lab, we run experiments that demonstrate the power of words to affect the brain. Our participants lie still in a brain scanner and listen to short descriptions of situations, like this one:

> You are driving home after staying out drinking all night. The long stretch of road in front of you seems to go on forever. You close your eyes for a moment. The car begins to skid. You jerk awake. You feel the steering wheel slip in your hands.

As our participants listen to these words, we see increased activity in regions of their brain that are involved in movement, even though their bodies are lying still. We see other activity in regions involved in vision, even though their eyes are closed. And here's the coolest part: there's also increased activity in the brain system that controls heart rate, breathing, metabolism, the immune system, hormones, and other internal gunk and junk . . . all from processing the meanings of words!

Why do the words you encounter have such wide-ranging effects inside you? Because many brain regions that process language *also control the insides of your body,* including major organs and systems that support your body budget. These brain regions, which are contained in what scientists call the "language network," guide your heart rate up and down. They adjust the glucose entering your bloodstream to fuel your cells. They change the flow of chemicals that support your immune system. The power of words is not a metaphor. It's in your brain wiring. We see similar wiring in other animals too; for example, neurons that are important for birdsong also control the organs of a bird's body.

Words, then, are tools for regulating human bodies. Other people's words have a direct effect on your brain activity and your bodily systems, and your words have that same effect on other people. Whether you intend that effect is irrelevant. It's how we're wired.

How far can these effects go? For example, can words be harmful to your health? In small doses, not really. When someone says things you don't like or insults you or even threatens your physical safety, you might feel awful as your body budget is taxed in that moment, but there's no physical damage to your brain or body. Your heart might race, your blood pressure might change, you might ooze sweat, and so forth, but then your body recovers and your brain might even be a bit stronger afterward. Evolution gifted you with a nervous system that can cope with these sorts of temporary metabolic changes and even benefit from them. Occasional stress can be like exercise. Brief withdrawals from your body budget followed by deposits create a stronger, better you.

But if you are stressed over and over and over again,

without much opportunity to recover, the effects can be far more grave. If you constantly struggle in a simmering sea of stress, and your body budget accrues an ever-deepening deficit, that's called chronic stress, and it does more than just make you miserable in the moment. Over time, *anything* that contributes to chronic stress can gradually eat away at your brain and cause illness in your body. This includes physical abuse, verbal aggression, social rejection, severe neglect, and the countless other creative ways that we social animals torment one another.

It's important to understand that the human brain doesn't seem to distinguish between different sources of chronic stress. If your body budget is already depleted by the circumstances of life — like physical illness, financial hardship, hormone surges, or simply not sleeping or exercising enough — your brain becomes more vulnerable to stress of all kinds. This includes the biological effects of words designed to threaten, bully, or torment you or people you care about. When your body budget is continually burdened, momentary stressors pile up, even the kind that you'd normally bounce back from quickly. It's like children jumping on a bed. The bed might withstand ten kids bouncing at the same time, but the eleventh one snaps the bed frame.

Simply put, a long period of chronic stress can harm a human brain. Scientific studies are absolutely clear on this point. When you're on the receiving end of ongoing insults and threats, for example, studies show that you're more likely to get sick. Scientists don't understand all the underlying mechanisms yet, but we know it happens.

These studies of verbal aggression tested average people across the political spectrum, left, right, and center. (We are all social animals, regardless of our stripes.) If people insult you, their words won't hurt your brain the first time, or the second, or maybe even the twentieth. But if you're exposed to verbal nastiness continually for months and months or if you live in an environment that persistently and relentlessly taxes your body budget, words can indeed physically injure your brain. Not because you're weak or a so-called snowflake, but because you're a human. Your nervous system is bound up with the behavior of other humans, for better or for worse. You can argue what the data means or if it's important, but it is what it is.

Two other studies, which I find remarkable as a scientist but unnerving as a person, measured the effects of stress on eating. One study found that if you're exposed to social stress within two hours of a meal, your body metabolizes the food in a way that adds 104 calories to the meal. If this happens daily, that's eleven pounds gained per year! Not only that, but if you eat healthful, unsaturated fats, such as those found in nuts, within one day of being stressed, your body metabolizes these foods as if they were filled with bad fats. I'm not saying this is a license to choose French fries over fish oil when you're stressed. You have to live with your own conscience there. But stress quite literally can make you gain weight.

The best thing for your nervous system is another human. The worst thing for your nervous system is also another human. This situation leads us to a fundamental dilemma of the human condition. Your brain needs other people in order to keep your body alive and healthy, and at the same time, many cultures strongly value individual rights and freedoms. Dependence and freedom are naturally in conflict. How, then, can we best respect and cultivate individual rights when we are social animals who regulate one another's nervous systems to survive?

To answer this question, I must loosen my white lab coat a bit as I gingerly dip a toe into political waters. There's an authentic tension between a belief in individual freedom, which implies you can say almost anything you want to anyone, and the biological fact that humans have socially dependent nervous systems, which means your words affect other people's bodies and brains. It is not a scientist's job to declare how to resolve this tension. But it is a scientist's job to point out that the biology is real and motivate people to grapple with the issues that play out in our social and political world. So here goes.

First off, any global solution to this dilemma is impossible, because different cultures have different values. Hate speech, for example, is legal in the United States as long as you don't overtly threaten to harm someone. In certain other parts of the world, simple criticism can get you a death sentence.

Moreover, in my experience, the fundamental dilemma of freedom versus dependence can be difficult to even discuss, let alone solve. If you attempt to have a dialogue about this dilemma in the United States, or even raise the issue, invariably someone will accuse you of being a socialist or claim that you're against the freedom of speech guaranteed by the First Amendment of the U.S. Constitution. Freedom, however, is a bipartisan issue around the world; we all want it, depending on the point in question. When debating gun ownership in the United States, conservatives tend to support personal freedom and liberals tend to advocate for control. When debating abortion, it's the other way around; conservatives tend to advocate control while liberals tend to support personal freedom.

Here in the United States, the solution to our dilemma is certainly *not* to restrict our freedom of speech. After all, history is filled with examples of overcoming our biology so we can live our values. Other people carry germs, for example, that can make us sick or even kill us, but only in the most nightmarish cases do we legislate a solution that restricts our personal freedoms. More commonly, we cooperate and innovate. We invent soap, we bump elbows instead of shaking hands, we search for new medicines and vaccines, and so on. If this is insufficient, experts tell us we're supposed to voluntarily isolate ourselves and practice social distancing. Even in a free society, our actions affect one another in ways that are, like viruses, often invisible to us.

A more realistic approach to our dilemma, I think, at least in the United States, is to realize that freedom always comes with responsibility. We are free to speak and act, but we are not free from the consequences of what we say and do. We might not care about those consequences, or we might not agree that those consequences are justified, but they nonetheless have costs that we all pay.

We pay the costs of increased health care for illnesses, like diabetes, cancer, depression, heart disease, and Alzheimer's disease, that are worsened by chronic stress. We pay the costs of ineffective government when politicians spew crap at one another and make personal attacks instead of having the reasoned debate that the Founding Fathers of the United States envisioned. We pay the costs of a citizenry that struggles to discuss politically charged topics with one another productively, a standoff that weakens our democracy.

We also pay the costs of reduced innovation in a global economy, because when people are persistently stressed, they don't learn as well. Creativity and innovation often mean failing repeatedly and having the tenacity to pick yourself up and try again. This extra effort takes extra energy. Your brain already burns 20 percent of your body's entire metabolic budget, making it the most "expensive" organ in your body, and every moment of your life, it makes economic decisions about what energy to spend, when to spend it, and when to save it. If you're already burdened with a body budget that's in the red, you're less likely to be a visionary spender.

Scientists are often asked to make their research useful to everyday life. These scientific findings about words, chronic stress, and disease are a perfect example. There is a real biological benefit when people treat one another with basic human dignity. And if we don't, there is also a real biological consequence, and it eventually trickles down to a financial and social cost for everyone. The price of personal freedom is personal responsibility for your impact on others. The wiring of all of our brains guarantees it.

As our society makes decisions about health care, the law, public policy, and education, we can ignore our socially dependent nervous systems, or we can take them seriously. These discussions may be difficult, but avoiding them is worse. Our biology won't just go away.

Taking our species' interdependence seriously doesn't mean restricting rights. It can mean simply understanding the impact we have on one another. Each of us can be the kind of person who makes more deposits into other people's body budgets than withdrawals or the kind of person who is a drain on the health and welfare of those around us.

Sometimes it's necessary to say things that other people find offensive or don't like. That's an essential part of democracy. But in these situations, do we just want to speak, or do we also want to be *heard*? If the latter, then our messages may be more effective if we give more consideration to how they're delivered. The form of delivery can make an already difficult message easier or harder on a listener's body budget. When we speak freely, it makes sense to communicate in ways that encourage others to listen.

Most people eat food farmed by others. Many live in homes built by others. Our nervous systems are tended by others. Your brain secretly works with other brains. This hidden cooperation keeps us healthy, so it matters how we treat one another in a very real, brain-wiring way. Therefore, we're not only more responsible for babies (lesson no. 3) and for ourselves (lesson no. 4) than we might think; we're also more responsible for other adults than we might think. Or want. Like it or not, we influence the brains and bodies of those around us with our actions and words, and they return the favor.

Lesson No.  6

Brains Make More than One Kind of Mind

WHEN PEOPLE FROM the island of Bali in Indonesia are afraid, they fall asleep. Or at least, that's what they're supposed to do.

Falling asleep might seem like a strange thing to do when you're afraid. If you're from a Western culture, you're supposed to freeze on the spot, widen your eyes, and gasp. You can also squeeze your eyes shut and scream, like a teenage babysitter in a bad horror movie. Or you can run away from whatever is scaring you. These behaviors are Western stereotypes for proper fear behavior. In Bali, the stereotype is to fall asleep.

What kind of mind snoozes out of fear? A kind of mind that's different from yours.

Human brains make many different kinds of minds. I don't just mean that your mind is different from your friends' and neighbors'. I'm talking about minds that have different basic features. For example, if you are from a Western culture, like I am, your mind has features called thoughts and emotions, and the two feel fundamentally different from each other. But people who grow up in Balinese culture, as well as in the Ilongot culture in the Philippines, do not experience what we Westerners call cognition and emotion as different kinds of events. They experience what we would call a blend of thinking and feeling, but to them it's a single thing. If you find this kind of mental feature hard to imagine, that's okay. You don't have a Balinese kind of mind.

As another example, Western minds often try to guess what other people are thinking or feeling. This mental inference is such a basic and valuable skill in our culture that when we encounter people who are not so good at it, we may see them as abnormal instead of merely different. But in some other cultures, attempts to peer into another person's mind are considered unnecessary. The Himba people of Namibia often figure each other out by observing each other's behavior, not by inferring a mental life behind that behavior. If you smile at an American, his brain might guess that you're happy to see him and predict that you'll say hello. If you smile at a Himba, his brain might predict only the hello (*moro*, in their language).

Even within a single culture, we find different kinds of minds. Think about the minds of great mathematicians who can conceive of calculations that other minds cannot. Or think about the mind of Greta Thunberg, a teenager who has sailed around the world offering tough talk about climate change. Thunberg's mind is on the autism spectrum, and she says things that others aren't willing to say. She calls her condition a "superpower" that helps her continue her mission when people criticize her efforts.

Think also about people who suffer from schizophrenia and experience severe, ongoing hallucinations. Today, people with this kind of mind are considered mentally ill, but centuries ago, they might have been called prophets or saints. Hildegard of Bingen, a twelfth-century scholar and nun, experienced visions of angels and demons and heard disembodied voices that were believed to come from God.

This variation in mind types should not, at this point in our lessons, come as a surprise. We have learned that humankind has a single brain architecture—a complex network—and yet each individual brain tunes and prunes itself to its surroundings. We've also learned that the mind and the body are strongly linked, and the boundary between the two is porous. Your brain's predictions prepare your body for action and then contribute to what you sense and otherwise experience.

In short, a particular human brain in a particular human body, raised and wired in a particular culture, will produce a particular kind of mind. There is not one human nature but many. A mind is something that emerges from a transaction between your brain and your body while they are surrounded by other brains-in-bodies that are immersed in a physical world and constructing a social world.

Let me be clear here. I'm not saying the human mind

is a blank slate and each of us becomes whatever the environment tells us to be, like there's nothing innate. That's the sort of mind that might emerge from Meatloaf Brain, the imaginary brain structure from lesson no. 2, in which every neuron is connected to every other. I'm also not saying people come into the world with their brains fully realized so that there's a single, universal human nature. That's the sort of mind that might emerge from Pocketknife Brain, the other imaginary brain structure, which consists of distinct brain regions, each with a dedicated function. I'm describing a third possibility. We come into the world with a basic brain plan that can be wired in a variety of ways to construct different kinds of minds.

It's important for humans to have many kinds of minds, because variation is critical for the survival of a species. One of Charles Darwin's greatest insights was that variation is a prerequisite for natural selection to work. Think about it: If there's a huge change in the environment, like a catastrophic drop in the food supply or a big increase in temperature, a species without much variation might be completely wiped out. A species with great variation will more likely have some survivors after any catastrophe—the members who are well suited for the new environment. Darwin observed variation in the bodies of animals, and the same principle applies to human minds. If we all had the same kind of mind—if there were only one human nature—then when disaster struck, we might become extinct. Thankfully, our species has many kinds of minds, both within a single culture and across cultures, so we're less likely to be wiped out. This variation preserves the evolvability of our species.

Even though variation is the norm—and a blessing for our species—it makes people uneasy. The idea of a single, universal human nature is so much more comfortable than continuous variation. So even when scientists do acknowledge that there are different kinds of minds, they try to tame the variation by organizing it into categories. They sort people into neat little boxes with labels. Some people are labeled as having a warm personality, and others are cold. Some people are more dominant and others more nurturing. Some cultures prioritize individuals over the group, while others do the opposite. Each box represents a feature of the mind that seems universal, and scientists use the boxes to catalog human minds.

You may have seen personality tests that collect information about you and assign you to a little box. A great example is the Myers-Briggs Type Indicator, or MBTI, which sorts people into sixteen little boxes labeled with different personality types to classify you and supposedly help you get ahead in your career. Sadly, the MBTI's scientific validity is pretty dubious. This test and its many cousins typically work by asking what you *believe* about yourself, which research suggests may have little to do with your actual behavior in daily life. Personally, I prefer the Hogwarts Sorting Test, which has only four boxes and is far more rigorous. (I'm a Ravenclaw.)

Scientists also try to organize the variation of minds by classifying what's normal and what's not. The problem is that "normal" is relative. For example, homosexuality was listed as a psychological illness for many years in the official catalog of mental disorders maintained by the American Psychiatric Association. Today, many people acknowledge a wide range of sexual orientations, identities, and genders as normal variation. (We're still cramming the variation into lots of little boxes, but it's a start.)

All of this organizing and labeling is an attempt to identify features of the mind that are universal across humankind. It seems like common sense that if you and I are part of the same species, along with a farmer in Buenos Aires, a shopkeeper in Tokyo, and a Himba goat-herder in Namibia, then all of these minds should be similar in certain ways. Some scientists even go looking for circuits in the brain that might house each so-called universal feature. And if they find a similar circuit in a nonhuman animal brain, they conclude that the animal has that psychological feature as well, and the world suddenly feels a bit cozier, like we've taken a step toward understanding evolved human nature.

But if there's one thing that's clear from our earlier lessons, it's that common sense isn't much use when it comes to understanding how a brain works. Brains have a lot of common features; minds, less so, because minds depend in part on micro-wiring that is tuned and pruned by culture. For example, many Western cultures draw strong dividing lines between the mental and the physical. If your stomach hurts, you're likely to visit your primary care physician or a gastroenterologist; if you're feeling anxiety, you're more likely to see a psychologist, even if the symptoms and the underlying causes are identical. But in some Eastern cultures, such as those that practice Buddhism, mind and body are much more integrated.

As far as I can tell, the human mind has no universal defining features. Pick any mental feature that's unique to humans, such as rich, spoken language, and you can always find some humans who don't have it, such as newborn infants. Alternatively, pick any mental feature that virtually all humans have, such as cooperation, and you can find plenty of other animals that have it too.

Even so, we can still find mental features that are widespread — because they're really, really useful, even if they aren't universal. One example is the ability to have relationships. It's useful to have a mind that defines itself in relation to others, particularly if your culture values the group over the individual. It's also useful to have a mind that separates itself from others, especially if your culture values the individual over the group. But people who care about neither themselves nor anyone else will have a hard time functioning in *any* human culture.

An especially useful feature of the mind, and one of the closest things we have to a universal mental feature, is mood — the general sense of feeling that comes from your body. Scientists call it *affect*.* Feelings of affect range from pleasant to unpleasant, from idle to activated. Affect is not emotion; your brain produces affect all the time, whether you're emotional or not and whether you notice it or not.

Affect is the source of all your joys and sorrows. It makes some things profound or sacred to you and other things trivial or vile. If you're a religious person, affect helps you feel connected to God. If you're a spiritual person but not necessarily religious, affect becomes the transcendent feeling of being part of something larger than yourself. If you're a skeptic, affect is what drives your certainty that others are wrong.

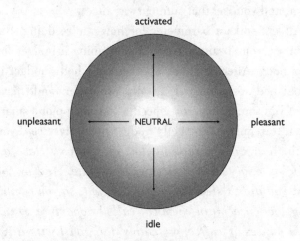

The properties of affect or mood

---

* The noun *affect* is pronounced with a short *a* as in *apple*, with the accent on the first syllable.

Where does affect come from? In every moment—
like right now, as you read these words—your hormones,
organs, and immune system are producing a storm of
sense data, and you're barely aware of it. You notice your
heartbeat and breathing only when they're intense or you
focus on them. You almost never notice your body temperature unless it's too high or too low. Your brain, however, makes meaning from this data storm continuously
to predict your body's next action and meet its metabolic
needs before they arise. In the midst of all this activity
inside you, something miraculous happens. Your brain
summarizes what's going on with your body in the moment, and you feel that summary as affect.

Affect is like a barometer for how you're doing. Remember, your brain is constantly running a budget for
your body. Affect hints at whether your body budget is
in balance or in the red. Ideally, evolution would have
given you something more specific, like an app or a smart
watch to regulate your body budget precisely. *Beep!* you
would hear. *You're running low on glucose. Have an
apple or, even better, a piece of chocolate. And by the
way, you didn't sleep enough last night, so you're running low on a brain chemical called dopamine. Drink
eight ounces of coffee, preferably dark roast with a little bit of cream, to borrow energy from tomorrow to get
through the rest of today.* Unfortunately, affect is not so
precise. It just tells you, *Beep! You feel like crap.* Then

your brain must predict what to do next to keep you alive and well.

Scientists are still puzzling out how your brain's body-budgeting activities, which are physical, become transformed into affect, which is mental. Hundreds of studies from laboratories around the world, including mine, observe that it happens, yet this transformation from physical signals to mental feelings remains one of the great mysteries of consciousness. It also reaffirms that your body is part of your mind — not in some gauzy, mystical way but in a tangible, biological way.

Even though every human culture produces minds that feel pleasure, displeasure, calmness, and agitation, we don't necessarily agree on *what* makes us feel these things. Some of us may find a gentle touch to be pleasant, others may find that same touch unbearable, and a few prefer a good spanking. Even here, variation is the norm. What the brain does to regulate the body may be universal, but the resulting mental experiences are not.

Your kind of mind is just one among many, and you are not stuck with the mind you have. You can modify your mind. People do this all the time. College students use caffeine or amphetamines to create minds that can pull an all-nighter before a final exam. Partygoers drink alcohol to create minds that are more relaxed and less inhibited in social situations (and miraculously, other people around them suddenly become much more attractive).

These chemical modifications last for only a short time. For longer-lasting modifications, you can try new experiences or learn new things to rewire your brain, as we discussed in earlier lessons.

A particularly challenging way to modify a mind is by moving it to another culture. If you've heard the story of the country mouse and the city mouse, or read *The Prince and the Pauper*, by Mark Twain, or seen movies like *Lost in Translation*, you know how it goes. The characters are thrust into cultures so unfamiliar that they don't know how to conduct themselves.

Imagine landing in a culture where you don't know even the most basic things. What is an acceptable way to greet people or even look at them? How close can you stand to other people without being rude? What do unfamiliar hand gestures and facial movements mean? Your mind must acclimate to the new culture. Scientists call this activity *acculturation*, and it's like an extreme version of plasticity. You're suddenly swimming in new and ambiguous sense data, and your brain needs to tune and prune itself so it can efficiently guess what to do.

Acculturation can be really challenging. If you've ever visited a country where people drive on the opposite side of the road, you know the mental pain of acculturation firsthand. Even the simple question of what is food and what is not food can be an adventure in a new culture. Imagine sitting down to eat and seeing for the first time an entire boiled sheep's head on your plate, or a bowlful of bee larva, or a Twinkie, for God's sake. One culture's food is another culture's inedible object.

Acculturation is not always about crossing geographic boundaries. You change cultures when you switch between work life and home life, and when you change jobs and have to learn the different norms and jargon of your new workplace. Military personnel have to acculturate at least twice—when they enter the armed forces *and* when they return home from deployment.

Your brain constantly issues predictions to manage your body budget, and if those predictions are out of sync with your current culture, your budget may accrue a deficit, which makes it easier for you to become sick. This is particularly true for the children of immigrants. They are of two cultures—their parents' culture and their adopted culture—and have to pivot between two kinds of minds, which adds a burden to their body budgets.

No kind of mind is inherently better or worse than any other. Some variations are just more tailored to their environment.

When it comes to human minds, variation is the norm, and what we call "human nature" is really many human natures. We don't need one universal mind in order to claim that we are all one species. All we need is an exceptionally complex brain that wires itself to its physical and social surroundings.

Lesson No.

Our Brains Can Create Reality

Most of your life takes place in a made-up world. You live in a city or town whose name and whose borders were made up by people. Your street address is spelled with letters and other symbols that were also made up by people. Every word in every book, including this one, uses those made-up symbols. You can acquire books and other goods with something called "money," which is represented by pieces of paper, metal, and plastic and is also completely made up. Sometimes money is invisible, flowing along cables between computer servers or traveling through the air as electromagnetic waves over a Wi-Fi network. You can even trade invisible money for invisible things, like the right to board an airplane early or the privilege of having another human serve you.

You actively and willingly participate in this made-up world every day. It is real to you. It's as real as your own name, which, by the way, was also made up by people.

We all live in a world of *social reality* that exists only inside our human brains. Nothing in physics or chemistry determines that you're leaving the United States and entering Canada, or that an expanse of water has certain fishing rights, or that a specific arc of the Earth's orbit around the sun is called January. These things are real to us anyway. Socially real.

The Earth itself, with its rocks and trees and deserts and oceans, is physical reality. Social reality means that we impose new functions on physical things, collectively. We agree, for example, that a particular chunk of Earth is a "country," and we agree that a particular human is its "leader," like a president or queen.

Social reality can alter dramatically, in moments, if people simply change their minds. In 1776, for example, a collection of thirteen British colonies vanished and was replaced by the United States of America. The world of social reality is also deadly serious. In the Middle East, people disagree and even kill each other over whether a parcel of land is Israel or Palestine. Even if we don't explicitly discuss the fact of social reality, our actions make it real.

The boundary between social reality and physical reality is porous, and we can use scientific experiments to reveal this. Studies show that wine tastes better when people believe it's expensive. Coffee labeled *ecofriendly* tastes better to people than identical, unlabeled coffee. Your brain's predictions, steeped in social reality, change the way you perceive what you eat and drink.

You and I can create social reality with other people without even trying, because we have human brains. To the best of our knowledge, no other animal brain can do that — social reality is a uniquely human ability. Scientists don't know for sure how our brains developed this capacity, but we suspect it has something to do with a suite of abilities that I'll call the Five Cs: creativity, communication, copying, cooperation, and compression.

First, we need a brain that's *creative*. The same creativity that permits us to make art and music also lets us draw a line in the dirt and call it the border of a country. This act requires us to invent some social reality (namely, countries) and impose new functions on an area of land, like citizenship and immigration, that don't exist in the physical world. Think about that the next time you pass through Customs, or even when you leave one town and enter another. Our borders are made up.

Next, we need a brain that can *communicate* efficiently with other brains in order to share ideas, such as the idea of a "country" and its "borders." Efficient communication for us usually includes language. For example, when I tell you that I need gas, I don't have to explain that I'm talking about my vehicle, not my digestive system, and that I plan to drive to a gas station in the near future, get out of my car, insert a plastic card into a pump for payment, and so forth. My brain conjures these features and so does yours, allowing us to communicate efficiently.

Strictly speaking, words are not necessary for social reality on a small scale. If your car and my car meet at an intersection and I wave for you to proceed first, you can observe my hand motion, guess its meaning, and use it yourself in the future. But for social reality to spread and persist, language is usually more efficient than other symbols. Imagine trying to establish and teach a country's driving laws without using words.

We also need brains that learn by reliably *copying* one another in order to establish laws and norms to live in harmony. We teach these norms to our children as we wire their little brains to their world. We teach them to newcomers, not only to smooth day-to-day interactions but also to help the newcomers survive. I've read about explorers in the 1800s who ventured into inhospitable, uncharted parts of the world, where many of them died. The expeditions that survived were the ones whose members became acquainted with the indigenous people in those regions; they taught the explorers what to eat, how to prepare the food, what to wear, and other secrets of survival in the unfamiliar climate. If all individual humans had to figure out everything themselves without copying, our species would be extinct.

We need brains that *cooperate* on a vast geographical scale. Even the most mundane act, like reaching into a kitchen cupboard for a can of beans, is possible only because of other humans. Other humans planted and watered those beans, perhaps thousands of miles away. Other humans mined the metal for the can. Still other humans transported the beans to your local store, which was built by other humans with wood and nails and bricks that were manufactured and hauled by other humans, using techniques and tools invented by other humans long dead. You paid for the beans with money that was invented and blessed by a government of other humans. Thanks to a shared social reality, all these thousands of people were in the right place at the right time doing the right things for you to grab the can and make dinner.

Creativity, communication, copying, and cooperation —four of the five Cs—arose with genetic changes that gave our species a big, complex brain. But large brain size and high complexity are not enough to make and maintain social reality. You also need the fifth C, *compression*, an intricate ability that humans have to a degree not found in any other animal brain. I'll explain compression first by analogy.

Imagine that you are a police detective investigating a crime by interviewing witnesses. You hear one witness's story, then another's, and so on, until you've interviewed twenty witnesses. Some of the stories have similarities —the same people involved or the same crime location. Some stories also have differences—who was at fault or what color the getaway car was. From this collection of stories, you can trim down the repetitive parts to create a summary of how the events might have occurred. Later, when the police chief asks you what happened, you can relay that summary efficiently.

A similar thing transpires among neurons in your brain. You might have a single, large neuron (the detective) receiving signals from umpteen little neurons at once (the witnesses) which are firing at various rates. The large neuron doesn't represent all of the signals from the smaller neurons. It summarizes them, or *compresses* them, by reducing redundancy. After compression, the large neuron can efficiently pass that summary to other neurons.

This neural process of compression runs at a massive scale throughout your brain. In your cerebral cortex, compression begins with small neurons that carry sense data from your eyes, ears, and other sense organs. Some of this data may already have been predicted by your brain, and some is new. The new sense data is passed by the small neurons to larger, better-connected neurons, which compress the data into summaries. Those summaries are passed to still larger, more highly connected neurons, which compress *those summaries* and pass them on to even larger, even more highly connected neurons. The process continues all the way to the densely wired front of your brain, where the very largest, most connected neurons create the most general, most compressed summaries of all.

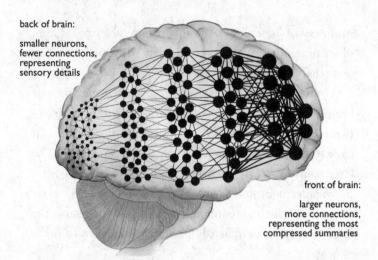

back of brain:

smaller neurons, fewer connections, representing sensory details

front of brain:

larger neurons, more connections, representing the most compressed summaries

Compression in the brain, which enables abstraction (this diagram is conceptual, not anatomically precise)

Okay, your brain can make a big, fat, compressed summary of summaries of summaries. What does this have to do with social reality? Well, compression makes it possible for your brain to think *abstractly*, and abstraction, together with the rest of the Five Cs, empowers your large, complex brain to create social reality.

Usually when people talk about abstraction, they mean something like abstract art, how you can look at a painting by Picasso and see a face in the cubes. Or they mean abstract math, like using algebra to rotate an object on its axes. Or they mean abstract symbols, like using a squiggle of ink on paper to represent a number, and a column of numbers to represent your spending for the month.

The psychological meaning of abstraction, though, has a different focus. It's not about the details of paintings and symbols; it's about our *ability* to perceive meaning in them. Specifically, we have the capacity to see things in terms of their function, not just their physical form. Abstraction lets you view objects that look nothing alike—such as a bottle of wine, a bouquet of flowers, and a gold wristwatch—and understand them all as "gifts that celebrate an achievement." Your brain compresses away the physical differences of these objects and in the process, you understand that they have a similar function.

Abstraction also allows you to impose multiple functions on the same physical object. A cup of wine means one thing when your friends shout, "Congratulations!" and another when a priest intones, "Blood of Christ."

Here's how abstraction works. As your brain compresses data from all your senses, it integrates them into a cohesive whole, an activity that we previously called sensory integration. Each time one of your neurons compresses its inputs to make a summary, that multisensory summary is an abstraction of the inputs. At the front of your brain, the largest, most highly connected neurons produce your most abstract, multisensory summaries. That's why you can view dissimilar objects like flowers and gold watches as similar and view an identical cup of wine as either celebratory or sacred.

I wrote in lesson no. 2 that you have a highly complex brain but high complexity isn't enough to make a human mind. Complexity may help you climb an unfamiliar staircase, but you need more to understand the idea of climbing a social ladder to gain power and influence. Abstraction is another necessary ingredient. It lets your brain summarize bits of past experience to understand that physically different things can be similar in other ways. Abstraction gives you the power to recognize things you've never encountered before, such as a woman with snakes for hair. You've probably never seen a real one, but you (and the ancient Greeks) could look at a picture of Medusa and instantly comprehend what she is, because, miraculously, your brain can assemble familiar ideas like *woman* and *wild hair* and *slithering snake* and *danger* into a coherent mental image. Abstraction also lets your brain assemble sounds into words, and words into ideas, so you can learn language.

In short: The wiring of your cerebral cortex makes compression possible. Compression enables sensory integration. Sensory integration enables abstraction. Abstraction permits your highly complex brain to issue flexible predictions based on the functions of things rather than on their physical form. That is creativity. And you can share these predictions by way of communication, cooperation, and copying. That is how the Five Cs empower a human brain to create and share social reality.

Each of the Five Cs is found in other animals to varying extents. Crows, for example, are creative problem-solvers who use twigs as tools. Elephants communicate in low rumbles that can travel for miles. Whales copy one another's songs. Ants cooperate to find food and defend their nest. Bees use abstraction as they wiggle their bums to tell their hive-mates where to find nectar.

In humans, however, the Five Cs intertwine and reinforce one another, which lets us take things to a whole other level. Songbirds learn their songs from adult tutors. Humans learn not only how to sing but also the social reality of singing, such as which songs are appropriate on holidays. Meerkats teach their offspring to kill by bringing them half-dead prey to practice on. We learn not only about killing but also the difference between accidental killing and murder, and we invent different legal penalties for each. Rats teach one another what's safe to eat by marking palatable foods with an odor. We learn not only what to eat but also which foods are main courses versus desserts in our culture and which utensils to use.

Other animals, such as dogs, great apes, and certain birds, also have brains that compress signals to a degree, so they can also understand things abstractly to some extent. But as far as we know, humans are the only animal whose brains have enough capacity for compression and abstraction to create social reality. A single dog might develop its own social rules, like that a particular grassy area is for playing with humans or that pooping is not allowed inside the house. But a dog brain cannot communicate these concepts to other dog brains efficiently the way human brains convey concepts with words to make social reality. Chimps can observe and copy one another's practices, like poking a stick into termite holes to pull out tasty snacks, but this learning is based in physical reality —namely, that sticks fit into termite holes. That's not social reality. If a troop of chimps agreed that whosoever pulls a particular stick out of the ground becomes king of the jungle, that would be social reality, because it imposes a sovereign function on the stick that goes beyond the physical.

Most animals have evolutionary adaptations that make them specialists in their niches, like an elk's antlers or an anteater's tongue. But humans became generalists; evolution blended the Five Cs into a potion that spurs us to bend the world to our will. All animal brains pay attention to things in their physical environment that are relevant to their well-being and survival and ignore the other stuff. But humans don't just select stuff from the physical world to create our niche. We *add* to the world by collectively imposing new functions, and we live by them. Social reality is human niche construction.

Social reality is an incredible gift. You can simply make stuff up, like a meme or a tradition or a law, and if other people treat it as real, it becomes real. Our social world is a buffer we build around the physical world. The author Lynda Barry writes, "We don't create a fantasy world to escape reality. We create it to be able to stay."

Social reality can also be a huge liability. It's so powerful that it can alter the speed and course of our genetic evolution. One example is the tragedy of the Romanian orphanages, when a government's rules created a generation of humans who were effectively removed from the gene pool. Another example is China's one-child policy, which, in a culture that values sons over daughters, led to more male offspring than female and ultimately to millions of Chinese men who cannot marry Chinese women. This sort of artificial selection happens in every society where wealth, social class, or war empowers one group over another — it changes the odds that certain people will reproduce with each other, or at all. Social reality even changes the course of human evolution when we simply share our creative ideas, such as the technology to burn fossil fuels, which has produced a physical world that is less under our control.

A really striking thing about social reality is that we often don't realize that we make it. The human brain misunderstands itself and mistakes social reality for physical reality, which can cause all sorts of problems. For example, humans vary tremendously, like every animal species does. But unlike the rest of the animal kingdom, we organize some of this variation into little boxes with labels such as race, gender, and nationality. We treat the labeled boxes as if they're part of nature when in fact we build them. Here's what I mean. The concept of "race" often includes physical characteristics such as skin tone. But skin tone is on a continuum, and boundaries between one set of shades and another are placed and maintained by people in a society. Some try to justify the boundaries by appealing to genetics, but while it's true that skin tone might be heavily influenced by genes, so are eye color, ear size, and the curvature of toenails. We, as a culture, choose the features of discrimination and draw dividing lines that magnify the differences between the group we call "us" and the group we call "them." The lines aren't random, but they aren't stipulated by biology either. And after the lines are drawn, people treat skin tone as a symbol for something else. *That* is social reality.

You uphold social reality by your everyday behaviors. You do it every time you treat sparkling diamonds like they have value, every time you idolize a celebrity, every time you vote in an election, and every time you *don't* vote in an election. Our behaviors can also change social reality. Sometimes the changes are relatively small, like using the pronoun *they* to refer to a single person instead of a group. Other times the changes are cataclysmic, like the breakup of the former country of Yugoslavia, which led to years of war and genocide, or the Great Recession of 2007, when some people in fancy suits decided that a bunch of mortgages had dropped in value, and so they did drop, plunging the world into catastrophe.

Social reality does have its limits; after all, it's constrained by physical reality. We could all agree that flapping our arms will let us soar into the air, but that won't make it happen. Even so, social reality is more malleable than you might think. People could agree that dinosaurs never existed, ignore all evidence to the contrary, and build a museum about a dinosaur-free past. We could have a leader who says terrible things, all captured on video, and then news outlets could agree that the words were never said. That's what happens in a totalitarian society. Social reality may be one of our greatest achievements but it's also a weapon we can wield against each other. It is vulnerable to being manipulated. Democracy itself is social reality.

Social reality is a superpower that emerges from an ensemble of human brains. It gives us the possibility to chart our own destiny and even influence the evolution of our species. We can make up abstract concepts, share them, weave them into a reality, and conquer just about any environment—natural, political, or social—as long as we work together. We have more control over reality than we might think. We also have more responsibility for reality than we might realize.

Every type of social reality is a dividing line. Some dividing lines help people, such as driving laws that prevent head-on collisions. Other dividing lines benefit some people and hurt others, such as slavery and social class. People debate the morality of such dividing lines, but like it or not, each of us bears some responsibility every time we reinforce them. A superpower works best when you know you have it.

Epilogue

ONCE UPON A TIME, you were a little stomach on a stick, floating in the sea. Little by little, you evolved. You grew sensory systems and learned that you were part of a bigger world. You grew bodily systems to navigate that world efficiently. And you grew a brain that ran a budget for your body. You learned to live in groups with all the other little brains-in-bodies. You crawled out of the water and onto land. And across the expanse of evolutionary time—with the innovation that comes from trial and error and the deaths of trillions of animals—you ended up with a human brain. A brain that can do so many impressive things but at the same time severely misunderstands itself.

➤ A brain that constructs such rich mental experiences that we feel like emotion and reason wrestle inside us

➤ A brain that's so complex that we describe it by metaphors and mistake them for knowledge

➤ A brain that's so skilled at rewiring itself that we think we're born with all sorts of things that we actually learn

➤ A brain that's so effective at hallucinating that we believe we see the world objectively, and so fast at predicting that we mistake our movements for reactions

➤ A brain that regulates other brains so invisibly that we presume we're independent of each other

➤ A brain that creates so many kinds of minds that we assume there's a single human nature to explain them all

➤ A brain that's so good at believing its own inventions that we mistake social reality for the natural world

We know much about the brain today, but there are still so many more lessons to learn. For now, at least, we've learned enough to sketch our brain's fantastical evolutionary journey and consider the implications for some of the most central and challenging aspects of our lives.

Our kind of brain isn't the biggest in the animal kingdom, and it's not the best in any objective sense. But it's ours. It's the source of our strengths and our foibles. It gives us our capacity to build civilizations and our capacity to tear down each other. It makes us simply, imperfectly, gloriously human.

Acknowledgments

This book owes its existence to many people, particularly the neuroscientists who educated me in their craft, guided my reading, and patiently answered my unending questions with unwavering generosity and good cheer. First and foremost is the incomparable Barbara Finlay. Barb is a connoisseur of evolutionary and developmental neuroscience. She regularly astounds me with her encyclopedic knowledge as she instructs me in the finer points of embryology and continually exposes me to a smorgasbord of neuroanatomy and neuroscience topics from the perspective of evolution and development. The half-lesson and lesson no. 1 in this book would not exist without Barb, and her fingerprints can be found in other lessons. Barb and I are currently collaborating on an academic book on the evolution and development of motivation and emotion in vertebrates, to be published by MIT Press.

I am also exceedingly grateful to my longtime collaborator and friend, neurologist Brad Dickerson. We've collaborated on brain-imaging studies for more than a decade at Massachusetts General Hospital in Boston, and we've published more than thirty research papers together. I particularly appreciate his willingness to indulge my sometimes exuberant scientific speculations. Special thanks also to Michael Numan, who was the first neuroscientist to encourage and support me as I began my neuroscience education.

My enduring thanks also go to my merry band of neuroscience collaborators not already mentioned, past and present, from whom I have learned so much. They include (in alphabetical order) Joe Andreano, Shir Atzil, Moshe Bar, Larry Barsalou, Marta Bianciardi, Kevin Bickart, Eliza Bliss-Moreau, Emery Brown, Jamie Bunce, Ciprian Catana, Lorena Chanes, Maximilien Chaumon, Sarah Dubrow, Wim van Duffel, Wei Gao, Talma Hendler, Martijn van den Heuvel, Jacob Hooker, Ben Hutchinson, Yuta Katsumi, Ian Kleckner, Phil Kragel, Aaron Kucyi, Kestas Kveraga, Kristen Lindquist, Dante Mantini, Helen Mayberg, Yoshiya Moriguchi, Suzanne Oosterwijk, Gal Raz, Carl Saab, Ajay Satpute, Lianne Scholtens, Kyle Simmons, Jordan Theriault, Alexandra Touroutoglou, Tor Wager, Larry Wald, Mariann Weierich, Christi Westlin, Susan Whitfield-Gabrieli, Christy Wilson-Mendenhall, and Jiahe Zhang. And I remain deeply grateful to my intrepid engineering and computer scientist collaborators, who continue to teach me about dynamical systems, complexity, and other topics in computation that make me a better neuroscientist, including Dana Brooks, Sarah Brown, Jaume Coll-Font, Jennifer Dy, Deniz Erdogmus, Zulqarnain Khan, Madhur Mangalam, Jan-Willem van de Meent, Sarah Ostadabbas, Misha Pavel, Sumientra Rampersad, Sebastian Ruf, Gene Tunik, Mathew Yarossi, and the rest of the PEN group at Northeastern University. Thanks also to statisticians Tim Johnson and Tom Nichols.

This book also wouldn't exist were it not for the boundless enthusiasm and expert guidance from my editor at Houghton Mifflin Harcourt, Alex Littlefield. I'm particularly grateful for his careful reading and his encouragement to combine complicated observations about the brain with big ideas of what it means to be a human being. In this regard, I'm also indebted to James Ryerson at the *New York Times* for his guidance as I developed my voice while navigating choppy waters between neuroscience, psychology, and philosophy.

The book also greatly benefited from the artistic skills and inquisitive nature of Van Yang, whose team's ingenious illustrations bring the science to life; I especially appreciate his deep desire to communicate science to a wide audience. Thanks also to Aaron Scott for his design consultations; his expertise, careful eye, and creativity have helped me translate complex scientific ideas into understandable images for over a decade.

Thank you to the production and marketing teams at

HMH, including Olivia Bartz, Chloe Foster, Tracy Roe, Chris Granniss, Emily Snyder, Heather Tamarkin, Taryn Roeder, Lisa McAuliffe, Hannah Dirgins, and especially Michelle Triant, PR maven extraordinaire. Thanks also to my agent, Max Brockman, for his continued enthusiasm and support, and to his crew at Brockman Inc., Thomas Delaney, Evelyn Chavez, Breana Swinehart, and Russell Weinberger.

This book was notably improved by valuable comments, criticisms, and ideas offered by early readers, many of whom are dear friends and extraordinary scientists in their own right. They are (in alphabetical order) Kevin Allison, Vanessa Kane Alves, Eliza Bliss-Moreau, Dana Brooks, Lindsey Drayton, Sarah Dubrow, Peter Farrar, Barb Finlay, Ludger Hartley, Katie Hoemann, Ben Hutchinson, Peggy Kalb, Tsiona Lida, Micah Kessel, Ann Kring, Batja Mesquita, Karen Quigley, Sebastian Ruf, Aaron Scott, Scott Sleek, Annie Temmink, Kelley Van Dilla, and Van Yang. And for close reviews of the science in specific lessons, I give special thanks to Olaf Sporns and Sebastian Ruf for lesson no. 2, Dima Amso for lesson no. 3, and Ben Hutchinson and Sarah Dubrow for lesson no. 4.

I also offer heartfelt thanks to my colleagues and trainees in the Interdisciplinary Affective Science Laboratory at Northeastern University and Massachusetts General Hospital. Much of the material in these essays has been the topic of ongoing discussion and research in our community of talented young scientists. All the members (past and present) are listed at affective-science.org. I'm particularly grateful to Sam Lyons for ultra-fast retrieval of a never-ending torrent of research papers on request and to Karen Quigley, who co-directs our lab. Karen has deep expertise in peripheral physiology of the body, interoception, and allostasis. We like to joke that, with her knowledge of the body and my knowledge of the brain, between the two of us, we make up a whole person.

I am also especially grateful to the Martinos Center for Biomedical Imaging at the Massachusetts General Hospital and its director, Bruce Rosen, as well as to the psychology department at Northeastern University, and in particular to our chair, Joanne Miller. Their support and patience make it possible for me to be both a neuroscientist and a psychologist, not to mention a communicator of science to the public.

This book was made possible with a fellowship from the John Simon Guggenheim Foundation and a book grant from the Alfred P. Sloan Foundation. I am deeply grateful to both for their generous support.

And above all, I offer a stream of continuous thanks and unbounded appreciation to the two brains I love best —my daughter, Sophia, and my husband, Dan—for their inspiration, forbearance, and general balancing of my body budget.

Appendix

The Science Behind the Science

This appendix adds crucial scientific details for certain topics in my essays, explains that certain points are still debated by scientists, and gives credit to scientists whose ideas and turns of phrase I've incorporated. Full references for the book can be found at sevenandahalflessons. com. (Most appendix entries also include a direct link to the relevant web page.)

The biggest challenge of science writing is deciding what to leave out. A science writer, like a sculptor, chips away at complex material until something compelling and comprehensible takes shape. The end result is necessarily incomplete from a strict scientific perspective, but (one hopes) still correct enough not to offend most experts.

An example of "correct enough" is saying that a human brain is made of approximately 128 billion neurons. This estimate may differ from some others that you've

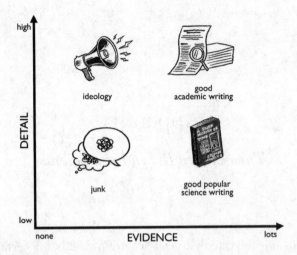

The challenge of writing about science for the public

seen, because I include the neurons that make up the cerebellum—a brain structure that's important for using sensations like touch and vision to coordinate physical movements, among other things. Some research papers may underestimate neurons in the cerebellum. Even so, my estimate of brain cells is incomplete, because the brain is also made of 69 billion other cells that are not neurons, called glial cells, which have a surprising number of biological functions. But the 128 billion figure serves to make the point that the brain is a complex network of parts, which is a pivotal concept in lesson no. 2.

Notes by page

The Half-Lesson: Your Brain Is Not for Thinking

1 *Amphioxi populated the oceans about 550 million years ago:* These ancient creatures, also called lancelets, are still around today. Amphioxi are our evolutionary cousins in the following way: Humans are vertebrates, meaning that we possess a backbone, which we call a spine, and a nerve cord, which we call a spinal cord. Amphioxi are not vertebrates, but they have a nerve cord running stem to stern. They also have a backbone of sorts, called a notochord, made of a fibrous material and muscle instead of bone. Amphioxi and vertebrates belong to a larger group of animals known as chordates (phylum Chordata), and we share a common ancestor. (More on this ancestor shortly.)

   Amphioxi lack all sorts of features that distinguish vertebrates from invertebrates. They have no heart, liver, pancreas, or kidneys, nor the internal bodily systems that go with these organs. They do have some cells that regulate a circadian rhythm and produce a cycle of sleeping and waking.

   Amphioxi do not have a distinct head or any of the visible sense organs that are found in a vertebrate head, such as eyes, ears, a nose, and so on. At its most anterior tip, an amphioxus has a small group of cells on one side, called an eyespot. These cells are photosensitive and can detect gross changes in light and dark, so if a shadow falls on the animal, the animal moves away. The cells of this eyespot share some genes in common with a vertebrate retina, but amphioxi do not have eyes and cannot see.

Also, amphioxi cannot smell or taste. They have some cells in their skin to detect chemicals in the water, and these cells contain some genes that are similar to those found in a vertebrate olfactory bulb, but it is not clear that the genes function in the same way. An amphioxus also has a cluster of cells with hairs in them that enable it to orient and balance its body in water and perhaps sense acceleration when it swims, but amphioxi do not have inner ears with hair cells to hear with, as vertebrates do.

Amphioxi also cannot locate food and approach it; they dine on whatever stream of little creatures the ocean currents deliver. They have cells to detect the *absence* of food and wriggle away in a random direction that hopefully will lead to a meal (in effect, the cells signal, *Anyplace is better than here*). See 7half.info/amphioxus.

1   *a teeny clump of cells that was not quite a brain:* Scientists continue to debate whether amphioxi have brains. It all comes down to where you draw the dividing line between "brain" and "not a brain." The evolutionary biologist Henry Gee sums up the situation well: "Nothing like the vertebrate brain is seen in either tunicates [sea squirts] or the amphioxus, although there are traces of its ground plan . . . if one looks hard enough."

Scientists pretty much agree that a sketch of the genetic outlines of the vertebrate brain can be found in anterior end of the amphioxus notochord, and these outlines are at least 550 million years old. This does not necessarily mean that the genes found in the anterior end of the notochord work in the same way or produce the same structures that they do in the brain of a vertebrate. (For more details on what it means for two species to have similar genes, see the appendix entry for lesson no. 1, "reptiles and nonhuman mammals have the same kinds of neurons that humans do.") And this is where the scientific debates begin. Amphioxi have some of the molecular patterns that organize the vertebrate brain into major segments, but scientists debate which segments are sketched out and which segment instructions are absent. It is also debatable whether the actual segments are present in amphioxi. Similarly, an amphioxus has the rudimentary genetic foundations necessary for a head, even though it has no head per se.

For a more detailed discussion of amphioxi, see Henry Gee's *Across the Bridge: Understanding the Origin of the Vertebrates*, and the evolutionary neuroscientists Georg Striedter and Glenn Northcutt's book *Brains Through Time: A Natural History of Vertebrates.* See 7half.info/amphioxus-brain.

2   *you behold a creature very similar to your own ancient, tiny ancestor:* Scientists believe that our common ancestor with amphioxi resembled modern amphioxi very closely, because amphioxi's environment (their niche) has barely changed in the past 550 million years, so they wouldn't have had to adapt much. In contrast, vertebrates have undergone tremendous evolutionary changes, as have other chordates, such as sea squirts. Therefore, scientists assume that by studying modern amphioxi, we can learn about the common ancestor of all chordates.

Still, some scientists continue to debate these assumptions—it's unlikely that amphioxi have not changed *at all* in half a billion years! For example, the amphioxus notochord (its central nervous system) extends the entire length of its body, from tip to tail, whereas in vertebrates, the spinal cord ends where the brain begins. Scientists debate whether our shared ancestor had an amphioxus-like notochord that became shorter in conjunction with evolving a vertebrate brain, or a shorter notochord that extended during evolution. Several similar debates (e.g., the evolution of olfaction) exist as well.

For a more detailed discussion of our amphioxus-like ancient ancestor, see Henry Gee's *Across the Bridge*. See 7half.info/ancestor.

2   *Why did a brain like yours evolve:* Statements like "Your brain is for *this*" and "Your brain evolved to do *that*" are examples of teleology, from the Greek word *telos*, meaning "end," "purpose," or "goal." Several types of teleology are discussed in science and philosophy. The most common type, which is generally discouraged by scientists and philosophers, is a statement that something was *intentionally designed for a purpose with an ultimate end point.* An example is suggesting that brains evolved in some kind of upward progression — say, from instinctual to rational, or from lower animals to higher animals. That is not the form of teleology I'm using in this lesson.

A second type of teleology, which I have employed in this lesson, is a statement that something is a *process that embodies a goal with no ultimate end point.* In stating that the brain is not for thinking but for regulating a body in a particular niche, I am not implying that body budgeting — allostasis — has some final end state. Allostasis is a process that anticipates and deals with ever-changing environmental input. All brains manage allostasis. There's no orderly progression from a worse way to a better way.

The psychologists Bethany Ojalehto, Sandra R. Waxman, and Douglas L. Medin study how people across cultures reason about the natural world. Their research suggests that teleological statements of the sort employed in this lesson reflect an appreciation of the relationships among living things and their environments. They call it "contextual, relational cognition." A statement like "A brain is not for thinking" is inherently relational (it refers to the relationship among the brain, various bodily systems, and stuff in the environment) and does not reflect that the brain was intentionally designed for a purpose with a final end point.

My phrasing (e.g., "Your brain is not for thinking") also appears in a particular context—in a nontechnical essay that describes aspects of brain function. The phrasing achieves its full meaning only in the context in which it's employed. If you strip away the context, it's easy to mistake the statement as the first, problematic type of teleology. Allostasis is of course not the sole cause of brain evolution and did not drive evolution in some orderly fashion. Brain evolution was largely driven by natural selection, which is haphazard and opportunistic. Brain evolution may also be influenced by cultural evolution, which I discuss in lesson no. 7. See 7half.info/teleology.

8   *The scientific name for body budgeting is* allostasis: Allostasis is not the only factor influencing how brains evolve and how they work, but it's a big one. Allostasis is a predictive balancing process over time, not a process that seeks a single, stable point for the body to maintain (it's not like a thermostat). The word for seeking a single, stable point is *homeostasis*. See 7half.info/allostasis.

8   *The movement should be* worth the effort, *economically
     speaking:* The idea of worthwhile movement is well stud-
     ied in the field of economics, where it's called *value.* See
     7half.info/value.

9   *the insides of bodies became more sophisticated:* The
     organs inside your body, such as your heart, stom-
     ach, and lungs, are called *viscera,* and they are part of
     broader visceral systems below your neck, such as your
     cardiovascular system, your gastrointestinal system, and
     your respiratory system, respectively. Movements that
     happen inside your heart, gut, lungs, and other organs
     are called *visceromotor* movements. Your brain controls
     your visceral systems (i.e., it performs visceromotor con-
     trol). In the same way that your brain has a primary mo-
     tor cortex and a whole system of structures in your sub-
     cortex for controlling your muscle movements, it also
     has a primary visceromotor cortex and a whole system
     of subcortical structures for controlling your viscera.
     Some visceral organs, like your lungs, require your brain
     in order to function. Your heart and your gut, however,
     have their own intrinsic rhythms, and the visceromo-
     tor system in your brain fine-tunes them. One last note:
     your body has other systems not typically linked to any
     visceral organ, such as the immune system and endo-
     crine system, and their changes are also broadly referred
     to as visceromotor.
         In the same way that the motor movements of your
     arms, legs, head, and torso produce sense data that is re-
     layed to your brain (specifically, to the somatosensory sys-
     tem), visceromotor movements produce sensory changes,
     called *interoceptive* sense data, that are sent to your brain

(to the interoceptive system). All this sense data helps your brain better control your motor and visceromotor movements.

The best scientific estimates today suggest that the evolution of visceral and visceromotor systems in vertebrates was accompanied by the evolution of sensory systems. After conception, when an embryo is building its brain and body, the visceral systems and the sensory systems both emerge from the same temporary cluster of cells, called the neural crest. So does the segment of the vertebrate brain that contains the visceromotor and interoceptive systems, which is known as the forebrain. The neural crest is unique to vertebrates and can be seen in all vertebrate species, including humans.

Visceromotor and interoceptive systems play a key role in determining the value of any movement, but we cannot say they evolved for that reason. Other selection pressures contributed to the evolution of the body's visceral systems and the brain's visceromotor system, such as the evolution of larger bodies that needed new kinds of tending and maintenance. For example, most animals on this planet are small in diameter, with only a few cells that span from the inside of the body to the outside world. This arrangement makes certain physiological functions easier, like the exchange of gases (in breathing) and removal of waste products. In a larger body, the inside of the body is farther away from the outside world, so new systems have evolved, like one that pumps water over gills to facilitate gas exchange, and the kidneys and an extended gut to excrete waste. These new systems allowed vertebrates to become more powerful swimmers and, accordingly, more successful predators. See 7half.info/visceral.

Lesson 1. You Have One Brain (Not Three)

13   *Your human mind, wrote Plato:* Plato wrote about the *psyche,* which differs from our modern idea of a mind. I am following the colloquial tradition of using *psyche* and *mind* as synonyms. See 7half.info/plato.

14   *scientists later mapped Plato's battle onto the brain:* The triune brain idea fused neuroscience with Plato's writings about the human psyche. In the early twentieth century, the physiologist Walter Cannon proposed that emotions were triggered and expressed (respectively) by two brain regions, the thalamus and the hypothalamus, which sit directly beneath the supposedly rational cortex. (Today, we know that the thalamus is the main gateway for all sense data, except for chemicals that become smells, to reach the cortex. The hypothalamus is critical for regulating blood pressure, heart rate, respiratory rate, sweating, and other physiological changes.) In the 1930s, the neuroanatomist James Papez proposed a "cortical circuit" dedicated to emotion. His circuit went beyond the thalamus and hypothalamus to include cortical regions that border the subcortical regions (the cingulate cortex) and were therefore assumed to be ancient. This segment of cortex was dubbed the limbic lobe by the neurologist Paul Broca fifty years earlier. (He used the term *limbic,* which comes from the Latin word meaning "border," *limbus.* This tissue abuts the brain's sensory systems and the motor system that moves your arms, legs, and other body parts. Broca thought the limbic lobe housed primitive survival faculties, like the sense of smell.) In the late 1940s, the neuroscientist Paul MacLean transformed Papez's "corti-

cal circuit" into a full-fledged limbic system and embedded it within a three-layered brain that he named the *triune brain*. See 7half.info/triune.

The outermost layer, part of the cerebral cortex: The many brain terms that include *cortex* can be confusing. The cerebral cortex is a sheet of neurons arranged in layers that covers the subcortical (meaning "below the cortex") parts of your brain. It is popularly believed that one part of the cerebral cortex is evolutionarily old and belongs to the limbic system (e.g., the cingulate cortex) and another part is evolutionarily new, which is why it's called the neocortex. This distinction derives from a misunderstanding of how the cortex evolved, which is the topic of this lesson.

15   *one of the most successful and widespread errors in all of science:* Scientists normally try to avoid saying that something is a fact or is definitively true or false. In the real world, facts have some probability of being true or false in a particular context. (As Henry Gee says in his book *The Accidental Species: Misunderstandings of Human Evolution,* science is a process of quantifying doubt.) In the case of the triune brain, however, it's justified to use more absolute language. By the time MacLean published his magnum opus, in 1990, *The Triune Brain in Evolution: Role in Paleocerebral Functions,* the evidence was already clear that the triune brain idea was wrong. Its continued popularity is an example of ideology rather than scientific inquiry. Scientists work hard to avoid ideology, but we are also people, and people are sometimes guided by belief more than data. (See Richard Lewontin's book

Biology as Ideology: The Doctrine of DNA.) Mistakes are part of the normal process of science, and when scientists acknowledge them, they are great opportunities for discovery. Learn more in Stuart Firestein's books *Failure: Why Science Is So Successful* and *Ignorance: How It Drives Science*. See 7half.info/triune-wrong.

18   *genes were most likely present in our last common ancestor:* This assumption depends on there not having been much evolutionary change in the cells of animals we're comparing.

More generally, genes are not the whole story when it comes to inferring whether two animals have brain features that can be traced back to a common ancestor even when those features look different to the naked eye. Sometimes genes can be misleading. And some scientists use other sources of biological information, such as the connections between neurons, to determine whether two brain structures have a common ancestry. For a more detailed discussion of this topic, which is called homology, see Georg Striedter's *Principles of Brain Evolution* and Striedter and Northcutt's *Brains Through Time*. See 7half.info/homology.

as brains become larger over evolutionary time, they reorganize: This idea comes from the neurobiologist Georg Striedter. He likened brains to companies, which reorganize to scale up their business. See Striedter's *Principles of Brain Evolution*. It is also possible for brains to lose complexity over evolutionary time or during development; an example is tunicates (sea squirts). See 7half.info/reorg.

19   *segregating and then integrating:* Here is an analogy to
reinforce my comparison of the primary somatosensory
cortex in rats and humans. The author and chef Thomas
Keller explains that if you cook a bunch of vegetables together in a pot, the mixture will have a single, blended
flavor. No individual ingredient stands out. But, Keller
explains, there's a better, tastier way to make your dish:
cook each vegetable separately and assemble them in the
pot at the end. Now every spoonful is a different complex
medley of flavors. The difference between these two techniques is essentially the difference between the primary
somatosensory cortex in rats and humans. The rat's single region is like a single pot containing all the ingredients, and the four human regions are like four pots with
separate ingredients. In the language of lesson no. 2, the
four-pot technique has higher complexity. See 7half.info/
keller.

*reptiles and nonhuman mammals have the same kinds
of neurons that humans do:* By this, I mean the neurons
have the same molecular identity — a specific gene or sequence of genes — that performs the same genetic activities (e.g., they make the same proteins). A given gene does
not necessarily make the same proteins in every animal
where it's found. Two animals can have the same genes,
but those genes can function differently or produce different structures. And even within the same animal, a network of genes can perform different genetic activities at
different times in development. (For a clear explanation
and examples, see Henry Gee's book *Across the Bridge.*)
The important observation here is that two creatures can
have neurons with some of the same genes that function

the same way in both creatures, and yet those neurons can differ in how they are organized, resulting in very different-looking brains. See 7half.info/same-neurons.

21 *The common brain-manufacturing plan:* This research originated with evolutionary and developmental neuroscientist Barbara Finlay, who calls it the "translating time" model. Finlay built a mathematical model that predicts the timing of 271 events in developing animal brains. Some of these events include when neurons are created, when axons begin to grow, when connectivity is established and refined, when myelin starts to form over the axons, and when brain volume starts to change and expand. Finlay's model calculates the equivalent number of days for any developmental event across eighteen mammalian species that have been studied and even some animal species not included in the original model. If one compares her model's predicted timing to the actual timing of brain formation, the correlation is an astounding 0.993 (on a scale of -1.0 to 1.0). This means the ordering of events is close to identical for all species studied, because they're all described by a single model.

Additionally, the genes found in various mammalian brain cells provide molecular genetic evidence that is consistent with the translating time model. The brain cells of jawed fish contain those genes as well. Some genes go all the way back to amphioxus and very likely to its common ancestor with humans. So, based on the genetic evidence alone, it's reasonable to infer that the common manufacturing plan (or part of it) holds for all jawed vertebrates. See 7half.info/manufacture.

22   *the human brain has no new parts:* As a neuroscientist,
     I am persuaded by the evidence that supports Finlay's
     hypothesis of a common brain-manufacturing plan. In-
     terested readers should be aware, however, that some sci-
     entists continue to hold to the idea that certain features
     of the human brain, such as the prefrontal cortex, have
     evolved to become larger than expected for a scaled-up
     primate brain. My view is that some of the distinctive ca-
     pacities of a human brain come from a combination of
     a big cerebral cortex (not bigger than expected for the
     overall brain size, mind you, just big in absolute terms)
     and souped-up connections between neurons in certain
     parts of the cortex, including upper layers of the prefron-
     tal cortex. Some scientists, myself included, hypothesize
     that these features give humans the ability to understand
     things by their function rather than their physical form,
     as I discuss in lesson no. 7 and in my earlier book, *How
     Emotions Are Made: The Secret Life of the Brain.* See
     7half.info/parts.

24   *There is no such thing as a limbic system dedicated to
     emotions:* Even though the limbic system is a myth, your
     brain does contain something called limbic circuitry. Neu-
     rons in limbic circuitry connect to the brain stem nuclei
     that regulate your autonomic nervous system, immune
     system, endocrine system, and other systems whose sense
     data create interoception, your brain's representation of
     the sensations in your body. Limbic circuitry is not exclu-
     sive to emotion and is distributed across multiple brain
     systems. It includes subcortical structures, such as the
     hypothalamus and the central nucleus of the amygdala;

allocortical structures, such as the hippocampus and the olfactory bulb; and parts of the cerebral cortex, such as the cingulate cortex and the anterior part of the insula. See 7half.info/limbic.

24 *The triune brain idea and its epic battle between emotion, instinct, and rationality is a modern myth:* The triune brain belongs to a long history of entrenched myths in science. Here are some more to amuse you. In the eighteenth century, serious scholars believed that heat was created by a mythical fluid called caloric and that combustion was caused by an imaginary substance called phlogiston. Physicists of the nineteenth century insisted that the universe was filled with an invisible substance called luminiferous ether that permitted light waves to propagate. Their medical colleagues attributed illnesses such as the plague to smelly vapors called miasmas. Each of these myths survived and substituted for scientific fact for one hundred years or more before it was overturned. See 7half.info/myths.

25 *we're just an interesting sort of animal:* This idea comes from Henry Gee's book *The Accidental Species.* See 7half.info/interesting.

Lesson 2. Your Brain Is a Network

30 *Your brain is a* network: Your brain network is made of smaller networks, or subnetworks, of interconnected neurons. Each subnetwork is a loose collection of neurons that constantly join in and leave as the subnetwork functions. Think of a basketball team that has twelve to fifteen

players but only five of them participate at a time. Players switch in and out of the game, but we still view the people on the court as the same team. Likewise, a subnetwork is maintained even though the actual neurons that create it switch in and out. This variability is an example of degeneracy, when structurally dissimilar elements (such as groups of neurons) perform the same function. See 7half .info/network.

31   *a network of 128 billion neurons:* My count of 128 billion neurons in the average human brain is higher than you may find in other sources, which commonly cite about 85 billion neurons. The difference is due to the fact that neurons can be counted by different methods. In general, scientists estimate the number of neurons in a brain using stereological methods, which employ probability and statistics to estimate the three-dimensional structure of neurons from two-dimensional images of brain tissue. The 128 billion figure comes from a paper that used a stereological method called optical fractionator that counted about 19 billion neurons in the human cerebrum, including the cerebral cortex, the hippocampus, and the olfactory bulb, and another 109 billion or so granule cells in the cerebellum, plus 28 million or so Purkinje neurons in the cerebellum. The more common figure of 85 billion neurons comes from another method called isotropic fractionator, which is simpler and quicker but systematically omits some neurons. See 7half.info/neurons.

A brain network is not a metaphor: The brain isn't symbolically *like* a network—it really *is* a network, meaning it functions similarly to other networks. The term *network* is a concept here, not a metaphor. It helps call to mind other networks that you know to help you understand better what a brain network is and how it works.

31 *Generally speaking, each neuron looks like a little tree:* The human brain has different types of neurons of various shapes and sizes. The kind of neuron I've described in our lesson is a pyramidal neuron in the cerebral cortex.

33 *I'll refer to this whole arrangement as the "wiring" of your brain:* The simple term *wiring*, as I use it, stands in for more specific structural details. In general, a neuron consists of a cell body, some branch-like structures on the top called dendrites (think the crown of a tree), and one long, slender projection with a root-like structure on the bottom called an axon. Each axon is much thinner than a human hair and has little balls on the end, called axon terminals, that are filled with chemicals. Dendrites are riddled with receptors to receive the chemicals. Typically, the axon terminals of one neuron are close to the dendrites of thousands of other neurons, but they do not touch, and the intervening spaces are called synapses. When a neuron's dendrites detect the presence of chemicals, the neuron "fires" by sending an electrical signal down its axon to its axon terminals, which release their neurotransmitters into the synapses; the neurotransmitters then attach to receptors on the other neurons' dendrites. (Other cells, called glial cells, help the process along and prevent chemical leaks.) This is how neurochemicals excite or inhibit the receiving neurons and change their rate of firing. Through this process, one individual neuron influences

thousands of others, and thousands of neurons can influence one, all simultaneously. This is the brain in action. See 7half.info/wiring.

37 *the area is routinely called the visual cortex:* What does it mean to "see"? Your conscious experience of things in the world, like seeing your hand or your phone, is created in part by neurons in your occipital cortex. It is possible to navigate the world if these neurons are damaged, however. If you place an obstacle in front of a person with damage to the primary visual cortex, the person won't consciously see the obstacle but will walk around it. This phenomenon is called blindsight. See 7half.info/blindsight.

if you blindfold people with typical vision: The study of blindfolded people who learned braille is another demonstration that neurons have multiple functions. When the scientists disrupted neural firing in the primary visual cortex (V1) using a technique called transcranial magnetic stimulation, blindfolded test subjects had a harder time reading braille, although that difficulty disappeared twenty-four hours after the blindfold was removed and visual input was available again to be processed by V1. See 7half.info/blindfold.

41 *A system has higher or lower complexity:* Complexity does not imply an orderly progression of brains on some phylogenetic scale or *scala naturae* from less complex to ever more complex, culminating in the human brain. The brains of other animals, such as monkeys and worms, also have complexity. See 7half.info/complexity.

41   *Meatloaf Brain:* I drew inspiration for this name from the book *The Blank Slate* by psychologist Steven Pinker; in it, he described a "uniform meatloaf" mind as "a homogeneous orb invested with unitary powers." See 7half .info/meatloaf.

42   *Pocketknife Brain:* This name was inspired by evolutionary psychologists Leda Cosmides and John Tooby, who described a human mind as like a Swiss Army knife. See 7half.info/pocketknife.

A real pocketknife with, say, fourteen tools: Here's a bit more mathematical detail behind the complexity of a fourteen-tool pocketknife. In a particular configuration of the pocketknife's tools, which I've called a pattern, each tool has two possible states: used or unused. Fourteen tools with two states each yields about 16,000 possible patterns for the whole pocketknife:

$$2\times2\times2\times2\times2\times2\times2\times2\times2\times2\times2\times2\times2\times2=2^{14}=16,384$$

Adding a fifteenth tool doubles the number of patterns:

$$2\times2\times2\times2\times2\times2\times2\times2\times2\times2\times2\times2\times2\times2\times2=2^{15}=32,768$$

If each tool is given an additional function, it now has three possible states instead of two—its first function, its second function, or unused. This yields far more total patterns for the pocketknife:

$$3\times3\times3\times3\times3\times3\times3\times3\times3\times3\times3\times3\times3\times3=3^{14}=4,782,969$$

Tools with four functions would yield 4^{14} or 268,435,456 patterns, and so on.

45   *Neurons aren't literally wired together:* This observation is courtesy of my colleague Dana Brooks in the Department of Electrical and Computer Engineering at Northeastern University.

Physicists sometimes say that light travels in waves: In this metaphor, I am not referring to wave-particle duality but to the myth of luminiferous ether described in an appendix entry in lesson no. 1. See 7half.info/wave.

Lesson 3. Little Brains Wire Themselves to Their World

47   *many newborn animals are more competent than newborn humans:* Of course, many newborn animals are less competent than newborn humans, such as the blind, bald little peanuts that are born to rats, guinea pigs, and other rodents.

51   *"Neurons that fire together, wire together":* This saying is attributed to the neuroscientist Donald Hebb, and the phenomenon is more formally known as Hebb's principle or Hebbian plasticity. Strictly speaking, the firing is not simultaneous—one neuron fires just before another. See Hebb's book *The Organization of Behavior: A Neuropsychological Theory.* See 7half.info/hebb.

54   *It has more of a lantern:* The wonderful metaphor of a "lantern of attention" is courtesy of psychologist Alison

Gopnik, who studies the cognitive development of children. See her book *The Philosophical Baby: What Children's Minds Tell Us About Truth, Love, and the Meaning of Life.*

Besides sharing attention, other abilities are probably important to developing a spotlight of attention. One is the brain's control of the head, an ability that develops over the first few months of life. Another is control of the muscles of the eye, called oculomotor control, which improves during the first few months of life.

I should also note that scientists still debate how much attentional capacity infants are born with and what kind of attentional capacities they might be. Many scientists who study development think that infants are genetically programmed to attend to certain features of the world (such as whether something is alive or not) and that subsequent development scaffolds onto these innate abilities. See 7half.info/lantern.

61   *it's far cheaper to eradicate poverty than to deal with its effects decades later:* Childhood poverty costs society close to one trillion dollars per year, according to a 2019 report by the National Academies of Sciences, Engineering, and Medicine, *A Roadmap to Reducing Child Poverty.* The cost of lifting children out of poverty, the report states, is far less than the price paid for the consequences of poverty after the children grow up. My colleague psychologist Isaiah Pickens points out the irony that in our culture, we start to treat people as more responsible for their actions right around the time that the ill effects of poverty and adversity manifest themselves in more serious ways. See 7half.info/poverty.

Lesson 4. Your Brain Predicts (Almost) Everything
You Do

64   *a man who served in the Rhodesian army:* Another take
on this story appears in my 2018 TEDx Talk "Cultivating Wisdom: The Power of Mood," which you can view at
7half.info/tedx.

66   *ambiguous scraps of sense data:* Sense data is not only
ambiguous but also incomplete. Information about the
world and the body is lost when it's processed by your
retina, cochlea, and other sensory organs and sent to the
brain. Scientists still debate just how much is lost, but everyone agrees that neurons convey less sense data from
the world and the body than is available to be perceived.
See 7half.info/incomplete.

Your brain assembles these bits into memories: The idea
that your brain uses past experiences to give incoming
sense data meaning is in some ways similar to immunolo-
gist and neuroscientist Gerald Edelman's proposal that
your ongoing conscious experience is the "remembered
present." See 7half.info/present.

69   *line drawings:* The three figures are a submarine going
over a waterfall, a spider doing a handstand, and a ski
jumper looking at spectators far below before pushing off.
     The figures are Droodles excerpted from *The Ulti-
mate Droodles Compendium — The Absurdly Complete
Collection of All the Classic Zany Creations of Roger
Price,* © 2019 Tallfellow Press, Inc. Used by permission.

Captions for the Droodles are: SUB-
MARINE GOING OVER A WATERFALL; SPIDER
DOING A HANDSTAND; SKI JUMP AND SPECTA-
TORS SEEN BY JUMPER. Tallfellow.com.

70   *"the beholder's share":* This idea about the perception of
artwork originated with the art historian Alois Riegl, who
called it "the beholder's involvement." The later term *beholder's share* was coined by art historian Ernst Gom-
brich. See 7half.info/art.

71   *an everyday kind of hallucination:* I referred to conscious
perception and experience as an everyday hallucination
for a number of years before discovering that philosopher
Andy Clark eloquently makes the same point, calling conscious experience a "controlled hallucination." See his
book *Surfing Uncertainty: Prediction, Action, and the
Embodied Mind.* Today, other scientists also describe experience in this way, notably the neuroscientist Anil Seth
in his engaging TED Talk "Your Brain Hallucinates Your
Conscious Reality." See 7half.info/hallucination.

81   *who bears responsibility when you behave badly:* Some
material on this topic comes from my 2018 TED Talk
"You Aren't at the Mercy of Your Emotions— Your Brain
Creates Them," which you can view at 7half.info/ted.

Lesson 5. Your Brain Secretly Works with Other Brains

89   *experiments that demonstrate the power of words:* My
      lab's research on the power of words, in which partici-
      pants listened to scenarios and imagined them while hav-
      ing their brain scanned, is discussed in several papers.
      See 7half.info/words.

      many brain regions that process language also control
      the insides of your body: The brain regions that scien-
      tists call the "language network" overlap to a large ex-
      tent with a network called the "default mode network,"
      particularly on the left side of the brain. The default
      mode network is part of a larger system that controls
      the internal systems of your body, including your au-
      tonomic nervous system (which controls your cardiovas-
      cular system, respiratory system, and other organ sys-
      tems), immune system, and endocrine system (which
      controls hormones and metabolism). See 7half.info/
      language-network.

91   *This includes physical abuse, verbal aggression:* Ver-
      bal aggression, at least the milder kind, depends on con-
      text. Not all profanity is verbal aggression. For example,
      women sometimes call each other *bitch* as a term of en-
      dearment or even empowerment. Likewise, words that are
      positive in one context can be aggressive in another. If
      you say something romantic to your partner who then re-
      sponds, "Come here and say that," your brain may pre-
      dict that a kiss is in your future. If you stand up to a bully
      who then responds, "Come here and say that," your brain
      may predict a threat. See 7half.info/aggression.

91   *a long period of chronic stress can harm a human brain:*
Studies show that chronic stress eats away at the brain
and the body over the long term regardless of whether the
stress stems from ongoing physical abuse, sexual abuse,
or verbal aggression. Scientific results like these are surprising and unwelcome, so it's helpful to consider the evidence in a bit of detail. I'll share just a small portion
here; more details are at 7half.info/chronic-stress.

First of all, chronic stress causes brain atrophy. It reduces brain tissue, notably in parts of the brain that are
important for body budgeting (allostasis), learning, and
cognitive flexibility.

What exactly causes atrophy in a stressed brain? And
how are these brain changes related to an increased likelihood of physical illness and a shorter life span? Scientists are still studying the biological details. One tricky
bit is that we can't view the microarchitecture of a living human brain in enough detail to know exactly what
changes occur. This is why scientists study the impact of
stress on nonhuman animals and then carefully generalize
to humans where possible. For example, see the research
of neuroendocrinologist Bruce McEwen.

Chronic verbal abuse in childhood has long-lasting effects. For example, in a study of 554 young adults, scientists asked the participants to rate their exposure to
verbal abuse from parents and peers when they were children. The scientists found that people who reported exposure to verbal abuse in childhood were more likely to
experience anxiety, depression, and anger during young
adulthood. Incredibly, these associations were larger than
those observed for people who reported physical abuse by
a family member and comparable to those observed for

people who reported sexual abuse by someone outside the family. These findings are consistent with the hypothesis that chronic verbal abuse in childhood predisposes people to mood disorders in young adulthood. However, an alternative interpretation is that people who suffer from mood disorders remember more abuse, including verbal abuse. That's why it's important to have other studies to help us determine which of these two hypotheses is more likely to be correct.

In one such study, scientists measured the biological impact of growing up in a harsh or chaotic family with a lot of verbal criticism and conflict. Researchers measured a marker of inflammation (interleukin 6) and a marker of metabolic dysfunction (cortisol resistance) in 135 female adolescents. Participants were interviewed four times during an eighteen-month period. Participants who reported a harsher family environment with more verbal aggression showed more immune dysfunction and more metabolic dysfunction as time went on, whereas participants with average exposure showed no change in these markers, and those with the lowest exposures were healthier. Other studies find similar results — swimming in a sea of sustained aggression places adolescents on a developmental trajectory that can lead to physical and mental illness.

An increasing number of studies consistently reveal a link between sustained social stress, usually involving verbal aggression, and an increased incidence of psychiatric and physical disease. For example, there is evidence that verbal aggression can alter the immune response sufficiently to reactivate latent herpes viruses, reduce the benefits of common vaccines, and slow the healing of wounds. These are not studies of vulnerable people but of average people who were drawn from across the political spectrum. I should also point out that these findings hold whether or not test subjects report *experiencing* intense stress. See 7half.info/chronic-stress.

92   *the effects of stress on eating:* I mentioned two studies on stress and how your body metabolizes food. Both studies are by psychologist Janice K. Kiecolt-Glaser and her colleagues. The figure of eleven pounds per year assumes that you're stressed before one meal each day — 104 calories times 365 days divided by 3,500 calories per pound. I like to offer up these scientific tidbits when I am at a flagging dinner party that needs a bit of livening up. See 7half.info/eat.

Lesson 6. Brains Make More than One Kind of Mind

98   *When people from the island of Bali in Indonesia are afraid, they fall asleep:* I borrowed this example from the psychologists Batja Mesquita and Nico Frijda. They cite an ethnology, *Balinese Character,* published in 1942, in which anthropologists Gregory Bateson and Margaret Mead observed that people who lived in Bali would often fall asleep when faced with events that were unfamiliar or frightening. Their interpretation was that the people were avoiding something scary, like you might do by closing your eyes during a gruesome or suspenseful movie. According to Bateson and Mead, sleeping was a socially approved response to fear; the Balinese called it *takoet poeles,* which translates to "in a fright sleep." See 7half .info/sleep.

100 *Thunberg's mind is on the autism spectrum:* Greta Thunberg describes herself as having Asperger's syndrome,
but the proper diagnostic term today is *autism spectrum
disorder.* See 7half.info/thunberg.

Hildegard of Bingen: Hildegarde of Bingen believed
that her visions, which she called "the Shade of the Living Light," were instructions from God. Over the years,
she documented her visions in words and artwork. Just to
be clear, I am *not* diagnosing Hildegard of Bingen with
schizophrenia or any other mental illness. Rather, I am
making a general point that one person's mystical experience can be another person's symptom of illness, depending on the historical or cultural context. A number of
scholars have retrospectively diagnosed Hildegard of Bingen with various disorders, but this sort of activity should
be done with extreme caution. See 7half.info/bingen.

101 *the sort of mind that might emerge from Pocketknife
Brain:* When applied to the mind (instead of the brain),
the clash of Pocketknife versus Meatloaf is perhaps best
known as nativism versus empiricism. This philosophical
debate is about whether knowledge is inborn or learned
from experience, and it has raged for thousands of years.
Psychologists sometimes call this debate faculty psychology versus associationism. See 7half.info/nativism.

variation is a prerequisite for natural selection to work:
In his book *On the Origin of Species,* Charles Darwin
proposed that variation among individuals in a species
is a prerequisite for natural selection during the course

of evolution. A species is a diverse group of individuals, and those who are most suited to a particular environment are more likely to survive and pass their genes to their offspring (who also will be more likely to survive and breed). Darwin's idea about variation, known as *population thinking*, is one of his greatest innovations, according to the evolutionary biologist Ernst Mayr. For a primer, see Mayr's book *What Makes Biology Unique*, and for a more thorough treatment, see his book *Toward a New Philosophy of Biology*. See 7half.info/variation.

102 *Myers-Briggs Type Indicator:* The MBTI and various other personality tests have no more scientific validity than horoscopes. Years of evidence show that the MBTI does not live up to its claims and does not consistently predict job performance. Nonetheless, these kinds of personality tests lure otherwise capable managers into making decisions that benefit neither their employees nor their company. Why do the test results seem so true when you receive them? Because the test asks what you *believe* about yourself. The results summarize those beliefs and give them back to you, and wow, they fit so well! The bottom line is this: You can't measure behavior by asking people their opinions about their behavior. You have to *observe* that behavior in multiple contexts. (Furthermore, the same people may be honest in some contexts and dishonest in others, introverted in some contexts and extroverted in others, and so on.) See 7half.info/mbti.

105 *Feelings of affect range from pleasant to unpleasant, from idle to activated:* Affect is described by a mathematical structure depicted in the figure on page 105,

called a *circumplex*, which was first discussed by the psychologist James A. Russell. A circumplex represents relations using the geometry of a circle; in this case, the relations among affective feelings. The term *circumplex* means "circular order of complexity" to indicate that the feelings in question are characterized simultaneously by at least two basic psychological features. The circle maps how similar the feelings are to one another, and the two dimensions describe the properties of similarity. See 7half .info/circumplex.

106 *an app or a smart watch to regulate your body budget:* This analogy also appears in my 2018 TEDx Talk "Cultivating Wisdom: The Power of Mood," which you can view at 7half.info/tedx2.

Lesson 7. Our Brains Can Create Reality

111 *The boundary between social reality and physical reality is porous:* This porous boundary is easily revealed by experiments about the sense of taste, such as the studies I mention in this lesson about wine and coffee. A more serious example can be found in lesson no. 3, where we discussed the vicious cycle of poverty. Societal attitudes toward people in poverty, which are social reality, affect the physical reality of brain development, which then increases the likelihood that those little brains will grow to become adults who live in poverty. See 7half.info/porous.

112 *a suite of abilities that I'll call the Five Cs:* The *Five Cs* is my own term for a collection of characteristics that evolve together to reinforce one another and that give humans the

capacity to create social reality on a large scale. Four of these Cs—creativity, communication, copying, and cooperation—are inspired by research from evolutionary biologist Kevin Laland, and my account draws heavily from his book *Darwin's Unfinished Symphony: How Culture Made the Human Mind*. Laland does not discuss the role of social reality in human evolution, but he discusses the related concept of cultural evolution. See 7half.info/5C.

113   *explorers in the 1800s:* The example of explorers who cooperated with indigenous people to survive comes from the anthropologist Joseph Henrich's book *The Secret of Our Success: How Culture Is Driving Human Evolution, Domesticating Our Species, and Making Us Smarter.* See 7half.info/explore.

114   *You also need the fifth C,* compression: Compression occurs in many parts of the brain. Here, we're discussing the compression that occurs in the cerebral cortex, particularly in layers 2 and 3. The human brain has souped-up wiring in these critical layers, which enhances compression.

A big, complex brain with the capacity to compress, however, is probably not sufficient on its own for small bits of social reality to cohere into a civilization. You also need the right metabolic conditions, including agriculture, to supply enough energy to build and maintain a human brain with its souped-up wiring. For a useful discussion, see Kevin Laland's book *Darwin's Unfinished Symphony*. Also see evolutionary biologist Richard Wrangham's book *Catching Fire: How Cooking Made Us Human.* See 7half.info/metabolic.

115 *sense data from your eyes, ears, and other sense organs:* Sense data is collected by various sense organs in your body, such as your eyes, ears, nose, and so on, and converted into neural signals that the brain can use. Sense data usually passes through several way stations before reaching the brain. For example, in vision, the cells in your retina (the thin layer that lines the back of your eyeball) are called photoreceptors and they convert light energy to neural signals. These neural signals travel along a bundle of nerve fibers called the optic nerve. A majority of your optic nerve fibers arrive at a cluster of neurons called the lateral geniculate nucleus, which is part of a brain structure called the thalamus; this structure's main job is to relay the sense data from your body and the surrounding world to your cerebral cortex. From there, the neural signals make their way to neurons at the very back of your cortex, in the occipital lobe, also known as your primary visual cortex. A small number of axons branch away from your optic nerve and travel to other parts of the subcortex, including your hypothalamus, which is a subcortical brain structure that is important for regulating the internal systems of your body.

Most of your sensory systems work in a similar way, except for the system that gives you your sense of smell, known as the olfactory system. The cells that convert chemicals in the air into neural signals are located in a structure called the olfactory bulb. These cells send information directly to the cerebral cortex, bypassing the thalamus. The neural signals bring olfactory sense data to your primary olfactory cortex, which is part of a brain region called the insula, which itself is a portion of the cerebral cortex between the temporal and frontal lobes. See 7half.info/sense-data.

116 *compression makes it possible for your brain to think* ab-stractly: Scientists are still working out the details of how the brain compresses information and how compression leads to abstraction. There is a long and vigorous debate about how much sensory and motor information remains in highly compressed abstractions. Some scientists pro-pose that abstractions are *multimodal*, meaning they in-clude information from all senses; others propose that abstractions are *amodal*, meaning they include no sense data. My view is that the evidence favors the multimodal hypothesis. For example, the most compressed summa-ries are created in areas of the cerebral cortex that neu-rologists and neuroanatomists call *heteromodal*, meaning that those areas manage information from multiple senses as well as motor information.

Presumably, a brain can achieve abstraction by other means than compression, because other animals without huge brains (such as dogs) or without a cerebral cortex (such as bees) can treat two things as similar based on their function—that is, they can do abstraction to some extent. See 7half.info/abstract.

119 *the Five Cs intertwine and reinforce one another:* This idea and its relevance to human evolution is the subject of an ongoing scientific debate. One evolutionary per-spective, known as the "modern synthesis," combines the science of genes (beginning with Mendelian genetics) and Darwin's theory of natural selection and assumes that genes are the only stable way to transmit informa-

tion from one generation to the next. An example would be the selfish-gene hypothesis by the evolutionary biologist Richard Dawkins. The other perspective, known as the "extended evolutionary synthesis," involves the various Cs and draws on findings that identify other sources of information transfer that are stable across generations (e.g., sense data from the visual environment that wires a brain during development, and the cultural transmission of information). The extended evolutionary synthesis, which considers evolutionary and developmental ("evo-devo") neuroscience, proposes other means of transfer, such as epigenetics and niche construction, as well as cultural evolution and gene-culture co-evolution. Examples are the views of Barbara Finlay and Kevin Laland. The breadth of this scientific debate is beyond the scope of our lessons here, but you can find a reading list at 7half .info/synthesis.

120 *it imposes a sovereign function on the stick that goes beyond the physical:* Chimpanzees and many other nonhuman animals have dominance hierarchies, but those hierarchies are neither established nor maintained by social reality. If every chimp in a troop agrees on which member is the alpha male, it is because the alpha may kill other animals who challenge him. Killing is physical reality. Most human leaders today stay in power without murdering their opponents. See 7half.info/sticks.

"We don't create a fantasy world to escape reality. We create it to be able to stay": This quote about fantasy worlds by author and cartoonist Lynda Barry comes from her book *What It Is.* See 7half.info/barry.

121   *physical characteristics such as skin tone:* Skin pigmentation has evolved and re-evolved in relation to the amount of ultraviolet light in the environment. Lighter skin tones are better adapted for environments with less ultraviolet (UV) light. Lighter pigmentation allows the skin to absorb more light and produce more more vitamin D, which is important for bone growth, bone strength, and a healthy immune system. In contrast, darker skin tones are better adapted for environments with more UV light, because darker pigmentation prevents the skin from absorbing too much light. This in turn slows the destruction of vitamin B_9, folic acid, which is important for cell growth and metabolism and is particularly important in early pregnancy (since sunlight breaks down folate). The intensity of UV rays is dictated by how close you are to the equator, but the amount of UV light that actually penetrates your skin depends on your skin pigmentation. A more detailed discussion can be found in anthropologist Nina Jablonski's book *Living Color: The Biological and Social Meaning of Skin Color.* See 7half.info/skin.

See more details at sevenandahalflessons.com!

Index